Modern Foundations of
Quantum Optics

Modern Foundations of
Quantum Optics

by

VLATKO VEDRAL
University of Leeds, UK

Imperial College Press

Published by

Imperial College Press
57 Shelton Street
Covent Garden
London WC2H 9HE

Distributed by

World Scientific Publishing Co. Pte. Ltd.
5 Toh Tuck Link, Singapore 596224
USA office: 27 Warren Street, Suite 401-402, Hackensack, NJ 07601
UK office: 57 Shelton Street, Covent Garden, London WC2H 9HE

British Library Cataloguing-in-Publication Data
A catalogue record for this book is available from the British Library.

First published 2005
Reprinted 2006

MODERN FOUNDATIONS OF QUANTUM OPTICS

ISBN-13 978-1-86094-531-1
ISBN-10 1-86094-531-7
ISBN-13 978-1-86094-553-3 (pbk)
ISBN-10 1-86094-553-8 (pbk)

Printed in Singapore

Dedicated to Ivona and Michael

Preface

This book represents the lecture notes for the course I gave at the Imperial College London for three years in a row between 2001 and 2004. I have edited the notes to make them more suitable for publication, but at the same time I have tried to change as little as possible in order to stay close to the spirit and style of the lectures which were an optional course for third and fourth year physics undergraduate studies. The course consisted of 26 lectures and three extra special topic lectures. The extra topics were intended to cover very recent advances in and applications of quantum optics. I focused on experiments on Rabi oscillations in cavity QED, on the achievement of atomic Bose–Einstein condensation and on quantum teleportation. These recent advancements — some of which have resulted in several recent Nobel prizes — show that quantum optics is a very exciting and important subject to learn.

The reader will see that in addition to the modern application, I have tried to present many topics in an original way, always keeping in mind modern developments and understanding. Of course, there are many standard derivations in my notes that can also be found in many other textbooks, some of them covered in much more detail in these other books. I pretend neither to have written a detailed nor a complete exposition of the subject. The choice of topics reflects very much my personal bias, my research interests and preferences. For example, I discuss the topic of Maxwell's demon and how the wave and particle nature of light can possibly be used to violate the second law of thermodynamics. I also discuss the notion of phase in quantum mechanics, the difference between dynamical and geometrical phases, as well as some very basic ideas behind the gauge principle and how electromagnetism can be derived from the Schrödinger equation. These additional topics, not traditionally covered by conventional texts, were intended to show that quantum optics is not an isolated subject, but that it is very intimately related to other areas of physics. They were also intended to break the monotony of the routine of only going through the, frequently tedious, background material. I wanted to show my students how exciting and lively the subject can be even at this introductory level, and that they can actively participate in it from the very start.

The order in which the notes are written is sometimes historical, sometimes didactic, frequently neither. More frequently than not they are written in the order of increasing complexity — which does not always coincide with the historical development. The logic of the course was to present different levels of our understanding of light — and quantum optics is the most sophisticated such understanding we have — through its interaction with matter. Loosely speaking, there are four levels in the notes: the classical, the old quantum, the semi-classical and the fully quantum level. I motivate some of the more traditional topics with examples that are both technologically and conceptually challenging. For example, I introduce the Mach–Zehnder interferometer with single photons at the very start to show not only that photons behave like particles and waves at the same time, but also that this can be exploited to perform operations that are unimaginable in classical physics — such as the interaction-free measurement. I have included five sets of problems and solutions. These are taken mainly from my three exam papers and are meant for the students to test their understanding of the presented material. Problem solving is, as always, crucial for understanding of any subject.

The notes end at the point where the field theory proper should begin. One could say — perhaps somewhat misleadingly — that quantum optics is the lowest order approximation to the full quantum field theory. From my experience in teaching, it seems that learning quantum optics first is a much better way of understanding the field theory than the usual second quantization formalism.

Finally, I had great fun working with students at Imperial College London, who not only taught me the subject, but also taught me how to teach. I hope you enjoy reading the notes as much as I enjoyed teaching the course!

V. Vedral

Acknowledgements

I would like to thank Artur Ekert for initially encouraging me to publish this book and for being supportive during the key stages of the publication process. The support and encouragement of Imperial College Press, especially Laurent Chaminade, is gratefully acknowledged.

I would like to thank all the third and fourth year students at Imperial College London between the years 2001 and 2004 for correcting many "typoes" and improving my notes a great deal by telling me what points need to be clarified. In particular thanks to William Irvine (now at Santa Barbara) for reading and revising a very early version of my notes (back in 2000). I am also grateful to Luke Rallan for his help with a very early version of the book.

I acknowledge Peter Knight, who proposed the first course on Quantum Optics at Imperial College London and whose syllabus I have modified only a bit here and there when I taught it myself.

Very special thanks goes to Caroline Rogers for preparing the manuscript for the final submission to Imperial College Press. She has redrawn many of the figures, as well as corrected and clarified some parts of the book. Her hard work was essential for the final preparation, which otherwise may have taken a much longer time to complete.

My deepest gratitude goes to my family, Ivona and Michael, who provide a constant source of inspiration and joy.

Contents

Chapter 1

From Geometry to the Quantum

According to one legend, Lucifer was God's favorite angel before stealing light from him and bringing it to mankind. For this, to us a generous act, Lucifer was expelled from heaven and subsequently became the top angel in hell. Most of us are not able to steal possessions from God, but we can at least admire his most marvellous creation — light. Quantum optics is the theory describing our most sophisticated understanding of light.

This book intends to acquaint you with the basic ideas of how physics describes the interaction of light and matter at three different levels: classical, semi-classical and quantum. You will be able to understand basic principles of laser operation leading to the ideas behind non-linear optics and multiphoton physics. You will also become familiar with the ideas of field quantization (not only the electromagnetic field, but also a more general one), nature of photons, and quantum fluctuations in light fields. These ideas will bring you to the forefront of current research. At the end of this book, I not only expect you to understand the basic methods in quantum optics, but also to be able to apply them in new situations — this is the key to true understanding. The notes contain five sets of problems, which are intended for your self-study. Being able to solve problems is definitely crucial for your understanding, and a great number of problems have been chosen from the past exam papers at Imperial College London set by me. I also hope — and this is I believe really very important — that the book will teach you to appreciate the way that science has developed within the last 100 years or so and the importance of the basic ideas in optics in relation to other ideas and concepts in science in general. The book contains a number of topics from thermodynamics, statistical mechanics and information theory that will illustrate that quantum optics is an integral part of a much larger body of scientific knowledge. I hope that at the end of it all, and this is really my main motivation, you will appreciate how quantum description of light forms an important part of our cultural heritage.

Optics itself is an ancient subject. Like any other branch of science, its roots can be found in Ancient Greece, and its development has always been inextricably linked to technological progress. The ancient Greeks had some rudimentary knowledge of geometrical optics, and knew of the laws of reflection and refraction, although they didn't have the appropriate mathematical formalism (trigonometry) to express these laws concisely. Optics was seen as a very useful subject by the Greeks: Archimedes was, for example, hired by the military men of the state to use mirrors and lenses to defend Syracuse (Sicily) by directing the Sun's rays at enemy ships in order to burn their sails. And like most of human activity (apart from some forms of art and mathematics) the Greek knowledge was frozen throughout the Middle Ages only to awaken more than 10 centuries later in the Renaissance. At the beginning of the 15th century, Leonardo da Vinci designed a great number of machines using light and was apparently the first person to record the phenomenon of *interference* — now so fundamental to our understanding not only of light, but matter too (as we will see later in this book). However, the first proper treatment of optics had to wait for the genius of Fermat and Newton (and, slightly later, Huygens) who studied the subject, making full use of mathematical rigor. It was then, in the 16th and 17th centuries, that optics became a mature science and an integral part of physics.

If you could shake a little magnet 428 trillion times per second, it would start making red light. This is not because the magnet would be getting hotter — the magnet could be cold and situated in the vacuum (so that there is no friction). This is because the electromagnetic field would be oscillating back and forth around the magnet which produces red light. If you could wiggle the magnet a bit faster, say 550 trillion times per second, it would glow green, while at around 800 trillion times per second it would produce light that is no longer visible — faster still and it would become ultraviolet. In the same respect, we can think of atoms and molecules as little magnets producing light — and their behavior as they do so is the subject of quantum optics.

From our modern perspective, optics can be divided into three distinct areas which are in order of increasing complexity and accuracy (they also follow the historical development):

- *Geometrical optics* is the kind of optics you would have done in your sixth form and the first year of university, prior to learning that light is an electromagnetic wave. Despite the fact that this is the lowest approximation of treating light, we can still derive some pretty fancy results with it — how lenses work, for instance, or why we see rainbows. I will assume that you are fully familiar with geometrical optics.

- *Physical optics* is based on the fact that light is an electromagnetic wave and, loosely speaking, contains geometrical optics as an approximation when the wavelength of light can be neglected ($\lambda \to 0$). Behavior of light as described by physical optics can be entirely deduced from Maxwell's equations, and it is this level of sophistication that we will investigate at the very beginning of the book.

- *Quantum optics* takes into account the fact that light is quantized in chunks of energy (called photons), and this theory is the most accurate way of treating light known to us today. It contains physical optics (and hence geometrical optics) as an approximation when the Planck constant can be neglected ($\hbar \to 0$). This treatment will be the core of the book.

Geometrical optics can be summarized in a small number of fundamental principles. For those of you interested in the colorful history of optics, I mention Huygens' *Treatise on Optics* as a good place to read about the early understanding of light. Here are the three basic principles that completely characterize all the phenomena in geometrical optics:

Geometrical Optics Principles

(1) In a homogeneous and uniform medium, light travels in a straight line.
(2) The angle of incidence is the same as that of reflection.
(3) The law of refraction is governed by the law of sines — to be detailed below (see Figure 1.1).

Are these laws independent of each other or can they be derived from a more fundamental principle? It turns out that they can be summarized in a very beautiful statement due to Fermat.

Fermat's Principle

Fermat's principle of least time. Light travels such that the time of travel is extremized (i.e. minimized or maximized).

All the above three laws can be derived from Fermat's principle. We will now briefly demonstrate this. The fact that in a homogeneous and uniform medium light travels in a straight line is simple, as the speed of light is the same everywhere in such a medium (by definition of the medium), and therefore a straight line, being the shortest path between two points, also leads to the shortest time of travel. The same reasoning applies for the incidence and reflection angles. The law of sines is a bit more complicated to derive, but I will now show you how to do so in a few lines. Suppose that light is going from a medium of refractive index 1 to a medium of refractive index n as shown in Figure 1.1.

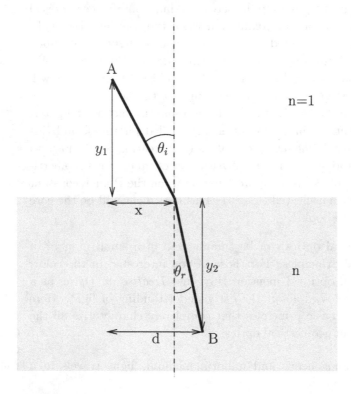

Fig. 1.1   The law of sines can be derived from Fermat's principle of least time. The full derivation is in the notes.

The total time taken from the point A to the point B is

$$t \propto \sqrt{x^2 + y_1^2} + n\sqrt{y_2^2 + (d - x)^2} \qquad (1.1)$$

Note that the second term is multiplied by n, as the speed of light is smaller in the medium of refractive index n, being equal to c/n where c is the speed of light in vacuum. Now, Fermat's principle requires that the time taken is extremized, leading to

$$\frac{dt}{dx} \propto \frac{x}{\sqrt{x^2 + y_1^2}} - \frac{(d - x)}{n\sqrt{y_2^2 + (d - x)^2}} = 0 \qquad (1.2)$$

which, after a short restructuring, gives

$$\sin \theta_i = n \sin \theta_r \qquad (1.3)$$

since $\sin \theta_i = x/\sqrt{x^2 + y_1^2}$ and $\sin \theta_r = (d - x)/n\sqrt{y_2^2 + (d - x)^2}$. Therefore, all three basic laws of geometrical optics can be derived from Fermat's least time principle. We can, of course, also ask "Why Fermat's principle?". But the reason for this cannot be found in geometrical optics. We need a more sophisticated theory to explain this.

Newton believed that light is made up of particles. Contrary to him, Huygens, who was his contemporary, believed that light is a wave. He reasoned as follows. If light is made up of particles then when we cross two different light beams, we would expect these particles to collide and produce some interesting effects. However, nothing like this really happens; in reality, the two beams just pass through each other and behave completely independently. The key property that in the end won the argument for Huygens against Newton was interference. That light exhibited interference was beautifully demonstrated by Young in his famous "double slit" experiment. Young basically observed a sinusoidal pattern of dark and light patterns (fringes) on a screen placed behind slits which were illuminated. The only way that this could have been explained was by assuming that light is a wave. However, the scientific community in England was not very favorable towards his findings and did not accept them for some time. Theoretically, the argument was clinched by Maxwell some 60 years after Young's experiment. He first came up with four equations fully describing the behavior of the electromagnetic field. These are the celebrated Maxwell's equations (I write their form in vacuum as this will be the relevant form for us here)

Maxwell's Equations

$$\nabla \mathbf{E} = 0 \qquad (1.4)$$

$$\nabla \wedge \mathbf{E} = -\frac{d\mathbf{B}}{dt} \qquad (1.5)$$

$$\nabla \mathbf{B} = 0 \qquad (1.6)$$

$$\nabla \wedge \mathbf{B} = \mu_0 \epsilon_0 \frac{d\mathbf{E}}{dt} \qquad (1.7)$$

where μ_0 is the permeability of free space and ϵ_0 is permittivity of free space. Maxwell was then very surprised to discover that he could derive a wave equation for the E and B fields propagating at the speed of light. This is very easy to obtain from the above equations (and you can find it in any textbook on electromagnetism):

we need to take a curl of the second equation and substitute the value of $\nabla \wedge \mathbf{B}$ from the last equation. We have

$$\nabla \wedge \nabla \wedge \mathbf{E} = -\nabla \wedge \frac{d\mathbf{B}}{dt} \qquad (1.8)$$

$$\nabla \wedge \mathbf{B} = \mu_0 \epsilon_0 \frac{d\mathbf{E}}{dt} \qquad (1.9)$$

which leads to the wave equation by using the fact that $\nabla \wedge \nabla \wedge =$ grad div $- \nabla^2$,

$$\nabla^2 \mathbf{E} = \frac{1}{c^2} \frac{\partial^2}{\partial t^2} \mathbf{E} \qquad (1.10)$$

where $c = 1/\sqrt{\mu_0 \epsilon_0}$ is the speed of light. The same wave equation can be derived for the magnetic field by manipulating the same two equations and reversing our steps (i.e. taking the curl of B first and then using the second equation). That this is so should be immediately clear from the symmetrical form of Maxwell's equations with respect to interchanging B and E. So, Maxwell concluded that light is an electromagnetic wave! Therefore, it displays all the wave properties: interference, in particular.

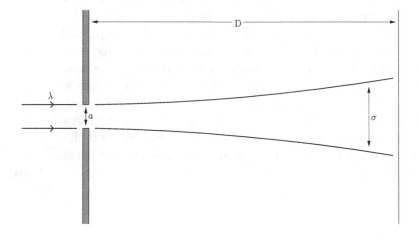

Fig. 1.2 Simple visualization of light diffraction. We observe in the laboratory that a light which passes through a small slit will spread in its width as it propagates. The distance beyond which the spread becomes significant (defined in the text) is called the Fraunhofer limit.

Let's describe a very simple interference behavior of a light beam of wave length λ, passing through a single slit of width a. A distance D after the slit we will obtain a bright spot of diameter σ. This spot will in general be larger than the size of the slit, which is the indication that light "bends around corners", i.e. it interferes.[1] There is a very simple relationship between the four quantities just mentioned which can be derived from a more rigorous wave optics treatment (see e.g. *Wave Optics* by Hecht):

$$\lambda D = \sigma a \qquad (1.11)$$

(Just think of an equation involving four numbers — dimensionally we have to multiply two numbers and equate them to the product of the other two. A logical way of doing so is to multiply the largest and the smallest number, D and λ respectively and equate them to the other two middle sized numbers — hence the above equation!) The Fraunhofer limit is the distance after which the light starts to spread, i.e. when $a = \sigma$. We therefore deduce

Fraunhofer Diffraction

$$D = \frac{a^2}{\lambda} \qquad (1.12)$$

This is a very useful formula to remember as it tells us under what conditions to expect light to start to behave like a wave (rather than travel in a straight line). Suppose that the slit is $1mm$ wide, and that $\lambda = 500nm$. Then for distances larger than $D = 4m$, light would behave like a wave. For distances below $4m$ light would for all practical purposes travel in a straight line — which is why geometrical optics is such a good approximation in the first place! (In a laboratory one would, of course, perform an interference experiment on a much smaller scale, and this would be achieved by putting a lens immediately after the slit to focus the light.)

What happens if light propagates not in the vacuum but in the air? Then there are atoms around which light can interact with. Imagine the following situation: a beam of light encounters two atoms as shown in Figure 1.3. The initial wave vector of light, which also determines the direction of propagation, is **k**. Suppose that the light changes its wave vector (and hence possibly the propagation direction as well) to **k**′ after scattering. Now I have to put you in the right frame of mind for calculating what we need from the wave formalism in order to show that light travels in straight lines. When we talk about waves, "amplitudes" become important. We have to add all the amplitudes for various possible ways that contribute to the process to obtain the total amplitude.[2] The final

---

[1] This behavior is strictly speaking called "diffraction", however, the fundamental process through which it arises is called interference, which is why I prefer to use this term. In fact, all the phenomena of light are just different consequences of the interference property.

[2] The fact that we have to add all the amplitudes is a consequence of the

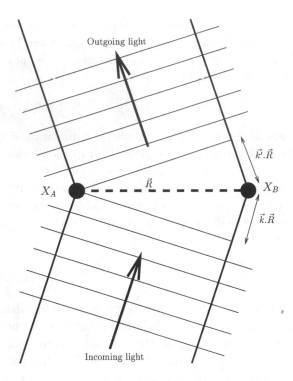

Fig. 1.3   Propagation of light in the air. We can derive the straight line trajectory from the wave theory of light.

total amplitude then has to be squared, leading to the intensity which is then the observable quantity. (Intensity is basically the number of photons falling onto a certain area per unit of time, but I don't really want to mention photons yet as we are not supposed to know quantum optics at this stage!) So, what is the final amplitude for this process? It is given by (strictly speaking, proportional to)

$$e^{\Delta \mathbf{k} \mathbf{x_A}} + e^{\Delta \mathbf{k} \mathbf{x_B}} = e^{\Delta \mathbf{k} \mathbf{x_B}} (1 + e^{\Delta \mathbf{k} \mathbf{R}}) \qquad (1.13)$$

where $\mathbf{x_A}$ and $\mathbf{x_B}$ are the position vectors of the two atoms (and $\mathbf{k}$ and $\mathbf{k}'$ are the initial and final light wave vectors respectively). So the intensity in the $\mathbf{k}'$-direction is given by the mod square of the amplitude

$$|1 + e^{\Delta \mathbf{k} \mathbf{R}}|^2 = 2(1 + \cos(\Delta \mathbf{k} \mathbf{R})) \qquad (1.14)$$

where $\Delta \mathbf{k} = \mathbf{k} - \mathbf{k}'$. Thus we see that if $\mathbf{k} = \mathbf{k}'$, then the intensity

linearity of wave equation; namely if two waves are solutions of this equation then so is their sum. We will talk about this in more detail later on.

is maximal, so according to Fermat's least time principle the light travels in a straight line. Of course, there will be other directions where we have maxima, given by $\Delta\mathbf{k} = 2n\pi$. So it looks as though light could take other paths than the straight line. However, imagine that there are more than two atoms, randomly distributed (like in the air, for instance, and *unlike* in periodic crystals as in a typical solid-state problem). Then any other direction will be unlikely as contributions from different $\mathbf{R}$s will average to zero unless the beam of light travels in a straight line. If it worries you that atoms are not moving in our treatment, just remember that the speed of light is typically 10^6 times larger. Thus, the first postulate of geometrical optics can be derived from the wave theory. With a little more effort it can be seen that the whole of geometrical optics can be derived as an approximation from Maxwell's equations! This reasoning is slightly simplified as light can also propagate in vacuum without any atoms around. The most general way of dealing with this is to take all the possible paths that light can take and add up all the corresponding amplitudes. The resulting amplitude should then be mod squared to yield the total intensity.

What changes in quantum optics? Well, light is again composed of particles (photons), but these particles behave like waves — they interfere (so both Newton and Huygens were somehow right after all). The proof for the existence of photons has built up over the year since Planck made his "quantum hypothesis" (which we will talk about in great detail shortly). I will mention a number of experiments throughout the book which demonstrate that light is composed of particles — photons. Now, however, I want to present a simple experiment to demonstrate the basic properties of quantum behavior of light. This is meant to motivate the rest of the subject without going into too much quantum mechanical detail at this stage.

The apparatus in Figure 1.4 is called the Mach–Zehnder interferometer. It consists of two beam splitters (half-silvered mirrors, which pass light with probability one half and reflect it with the same probability), and two 100 percent reflecting mirrors. Let us now calculate what happens in this set up to a single photon that enters the interferometer. For this we need to know the action of a beam splitter. The action of a beam splitter on the state a is given by the simple rule

$$|a\rangle \rightarrow |b\rangle + i|c\rangle \tag{1.15}$$

Beam Splitter Transformation

which means that the state a goes into an equal superposition of

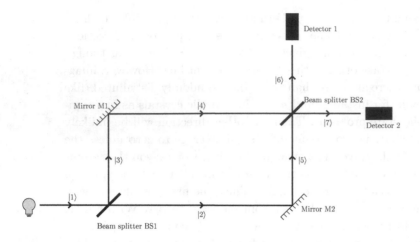

Fig. 1.4   Mach–Zehnder Interferometer. This is one of the most frequently used interferometers in the spectral study of light. In this book we will use it mainly to illustrate the unusual behavior of light in quantum mechanics.

states b and c.[3] The imaginary phase in front of c signifies that when light is reflected from a mirror at 90° it picks up a phase of $e^{i\pi/2} = i$ (the origin of this phase is purely classical, that is, derivable from Maxwell's equations). Now the Mach–Zehnder interferometer works as follows:

Quantum Interference

$$|1\rangle \overset{BS1}{\to} |2\rangle + i|3\rangle \overset{M1,M2}{\to} (i|5\rangle) + i(i|4\rangle) \tag{1.16}$$

$$= i|5\rangle - |4\rangle \overset{BS2}{\to} i(|6\rangle + i|7\rangle) - (i|6\rangle + |7\rangle) \tag{1.17}$$

$$= i|6\rangle - |7\rangle - i|6\rangle - |7\rangle = -2|7\rangle \tag{1.18}$$

Therefore, if everything is set up properly, and if both of the arms of the interferometer have the same length, then the light will come out and be detected by detector 2 only.[4] This is called interference and is a well-known property of waves, as we saw (it's just that quantumly every photon behaves in this way). What would happen if we detected light after the first beam splitter and wanted to know which route it took? Then, half of the photons would be detected in arm 2 and the rest of them would be detected in arm 3. So, it seems that photons randomly choose to move left or right at a beam

---

[3]Note that this state is not normalized. We need a prefactor of $1/\sqrt{2}$, but since the normalization is the same for both states b and c we will omit it throughout.

[4]Because we did not normalize the initial state and the states throughout the interferometer, there is an extra factor of "2" in the final result which should be ignored. The extra minus sign is just an overall phase and cannot be detected by any experiment.

splitter. And they are particles; we never detect half the photon in one arm and the other half in the other arm — they come in chunks. Thus it seems that this is the same as tossing a coin and registering heads or tails. Well, not quite. In fact, not at all. The interferometer shows why. Suppose that at the first beam splitter the photons goes either left or right, but it definitely goes either left or right (as our experience seems to suggest). Then, at the second beam splitter the photon would again face the same choice, i.e. it would definitely move either left or right. So, according to this reasoning we should expect detectors 1 and 2 to click with an equal frequency. But this is not what we saw! In reality, only detector 2 clicks. The amplitude, and therefore the probability for detector 1 to click is zero. This means that the operation of a beam splitter and the behavior of the photon is not just like coin tossing. The state after the beam splitter is more than just a statistical (random) mixture of the two probabilities. It is, of course, a *superposition*, and the photon takes both of the possible routes (in spite of being a particle). This is the true meaning behind writing its state as a mathematical sum of two vectors, say $|1\rangle + |2\rangle$.[5] This is basically why we use vectors to express states of physical systems.

But we won't be using this most sophisticated description of light immediately from the beginning of the book. Why? There are several reasons for this. Firstly, the mathematics used for the full quantum theory of light is quite advanced. Secondly, there are many important features of light that can be correctly described using the less sophisticated (so called) semi-classical theory. So we don't need to bother with the more complicated manipulations of equations and can postpone this until later. Thirdly, starting with simple things and going towards more complicated stuff has a great pedagogical value. It shows us how our understanding improves and teaches us never to be dogmatic about our understanding since it is very likely that it will be superseded by some better theory. I'd say that this is the most important part of our scientific culture. And finally, if we started with the complicated theory we would miss out on all the beautiful progress that took place at the beginning of the last century, and it was precisely this progress that made it the Century of Physics. Therefore, in the first part of this book (up to Chapter 7) we will have to do a lot of "hand waving" in order to describe the interaction of light with

---

[5] So photons (and all other physical systems as it turns out) behave according to Yogi Berra's saying: "When you come to a fork in the road, take it!"

matter, which will only be justified by a rigorous quantum treatment in the second part. However, we are in good company, as this is exactly what Planck and Einstein had to do about 100 years ago!

Final Thought: What have we learnt from the above story? Optics is an old science and the story of light has evolved over many centuries. The first "modern" treatment of light (Newton's *Optics* in 1660) described light as composed of small particles — corpuscules. This was compatible with the fact that light travels in straight lines, but there were phenomena difficult to explain, such as interference. With the discovery of Maxwell's equations, it was firmly established that light was an electromagnetic wave; therefore it interferes and diffracts. However, this theory was also found not to be completely compatible with some experimental evidence, in particular the Compton scattering as we will see later on in the book. Finally, quantum mechanics united the previous two and combined them into a new picture where light is composed of particles which interfere at the individual level. Science thus produces better and better approximations of nature to account for the more thorough experimental evidence that we gather through more developing technology. With every new scientific theory our understanding and picture of the world change dramatically and usually result in a different philosophy. It will not be surprising at all if all the results presented in this book are superseded by a higher level generalization of which they become an approximation in the same way that today classical optics approximates quantum optics. (This, of course, doesn't mean that what we will learn in this book would become useless; on the contrary, it will become crucial in testing quantum mechanics and exploring its domain of validity and applicability.)

Chapter 2

Introduction to Lasers

We motivate this section with a very bizarre consequence of quantum mechanics, called the interaction-free measurement,[1] which has been performed experimentally using lasers (and beam splitters).[2] Suppose that in the Mach–Zehnder interferometer we block one of the paths after the first beam splitter, say path 5, by inserting an absorbing material as shown in Figure 2.1.

Interaction-Free Measurement

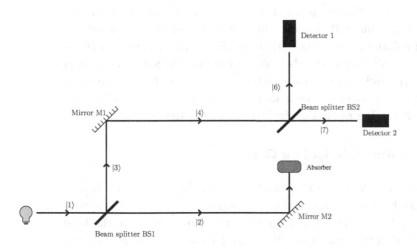

Fig. 2.1   This is the set up involving a Mach–Zehnder interferometer which shows how strange quantum mechanics is, and exemplifies the weird behavior of quantum objects. The presence or absence of the absorber can be determined without interacting with it! This leads to the notion of the interaction-free measurement that is a hot subject of current research into quantum mechanics.

---

[1]This notion was introduced by Elitzur and Vaidman, Found. Phys. **23**, 987 (1993).

[2]The experiment was performed by Kwiat *et. al*, Phys. Rev. Lett. **74**, 4763 (1995). You can read much more about interaction-free measurement in Penrose's semi-popular book [Penrose (1993)].

What happens when the absorbing material is inserted? Well if the photon is absorbed, then neither of the two detectors will eventually "click" — that's fine. However, if the photon takes the other path, then at the last beam splitter it has an equal chance to be reflected and transmitted so that the two detectors click with equal frequencies — in other words, the interference has been destroyed by the presence of the absorber in path 5. But, here is a very weird conclusion: we can detect the presence of an absorber in path 5, without the photon even being absorbed by it — hence interaction-free measurement! If detector 1 clicks, then the photon has gone to path 6, and that implies that there is an obstacle in path 5, or else only detector 2 would be clicking. This is surely amazing! But that is the basis of the quantum mechanical description of light and all the wonderful phenomena we'll be talking about in this book.[3]

The invention and development of lasers has been paramount to understanding the interaction between light and matter and has lead to a plethora of applications (we will cover a great deal of those, among which are two that lead to two recent Nobel Prizes in physics). "Laser" is short for Light Amplification by Stimulated Emission of Radiation (the fact that radiation emission can be stimulated is one of the surprising discoveries of Einstein's treatment we will discuss shortly). We will see what this means in great detail throughout this book. Also, some of the most exciting applications of physics come from the use of lasers — we will discuss a small number of "hot topics" at the end of the book.

2.1   Normal Modes in a Cavity

A laser consists of a cavity with a certain lasing medium inside. Physically, it is just a bunch of atoms oscillating inside a box with highly reflecting mirrors. I would first like to talk about the kind of radiation that we expect when we have a set up like this. For that matter, let's simplify the situation even further. Let's imagine that there are no atoms inside, just a cavity with highly reflecting mirrors. What do Maxwell's equations tell us about the radiation inside? The simplest solution of Maxwell's equations is the plane wave of the form

---

[3]This story can be made much more dramatic by imagining that instead of the absorber we have a box which may or may not contain a superbomb. This bomb is so sensitive that it explodes if a single photon hits it. So to check if the box is hiding the bomb or not we cannot lift it as the bomb will then be illuminated and hence will explode destroying the world. Here the Mach–Zehnder set up and the interaction-free measurement come to rescue.

$$\mathbf{E} = \mathbf{E_0}e^{i(\mathbf{kr}-\omega t)} \qquad (2.1)$$

where all the symbols have their usual meaning. But light in a cavity cannot be a free-propagating wave. First of all, we have to see what happens at the walls of the cavity. Since they are highly reflecting mirrors, this means that the electric field on the surface is very nearly zero (there is some penetration into the mirror, but this can be neglected) as shown in Figure 2.2.

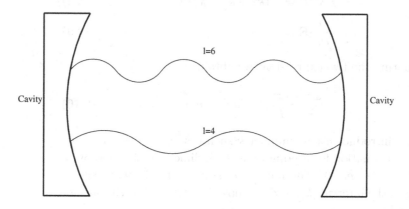

Fig. 2.2 Modes in a cavity are defined by their wavelengths. Wavelengths can only assume certain sizes such that an integral number of half wavelength is equal to the length of the cavity. This is a consequence of the fact that the electric field has to disappear at the walls of the cavity.

With this boundary condition we can write the following form of the field inside the cavity:

$$E_x(t) = E_{0x}(t) \cos k_x x \sin k_y y \sin k_z z \qquad (2.2)$$

$$E_y(t) = E_{0y}(t) \sin k_x x \cos k_y y \sin k_z z \qquad (2.3)$$

$$E_z(t) = E_{0z}(t) \sin k_x x \sin k_y y \cos k_z z \qquad (2.4)$$

where $k^2 = k_x^2 + k_y^2 + k_z^2$, and

$$k_x = \frac{l\pi}{L} \qquad (2.5)$$

$$k_y = \frac{m\pi}{L} \qquad (2.6)$$

$$k_z = \frac{n\pi}{L} \qquad (2.7)$$

where l, n and m are integers and L is the length of the cavity in each of the three directions. So, the wave vector comes in each direction in discrete units of π/L. The equations (2.6), (2.7) and

(2.7) are needed to ensure that the electric field disappears at the walls of the cavity. This resulting state can be checked to satisfy the wave equation (do it!) and, in fact, any superposition of these basic waves also does (this is because the wave equation is linear). But there is also a temporal dependence of the field of the form $e^{\pm i\omega t}$ which we didn't write down explicitly. Plugging this into the wave equation we see that only certain frequencies are allowed. This is because

$$\nabla^2 \mathbf{E} = -(k_x^2 + k_y^2 + k_z^2)E \qquad (2.8)$$

$$\frac{\partial^2}{\partial t^2} \mathbf{E} = -\omega^2 E \qquad (2.9)$$

By equating the two expressions we obtain

$$\frac{\omega^2}{c^2} = \frac{\pi^2}{L^2}(l^2 + m^2 + n^2) \qquad (2.10)$$

Let's introduce a new number k such that $k^2 = l^2 + m^2 + n^2$. This is an equation for a point on a three-dimensional sphere with co-ordinates (l, m, n). The number of states with the wave vector k lying in the interval $(k, k+dk)$ is now proportional to the surface of the sphere so that

$$n(k)dk = 4\pi k^2 dk \qquad (2.11)$$

(This is the volume whose area is $4\pi k^2$, thickness dk). This is, however, a continuous number (which cannot be true as we saw that the wave vector is discretized) and also has incorrect dimensions (it should be dimensionless as it refers to the *number* of states or modes). We have seen that, contrary to this, the wavelength, and hence wave vector, and momentum come in discrete units. The wave vector comes in units of π/L in every space direction. Thus, we obtain the following expression for the number of states (modes)

$$n(k) = \frac{4\pi k^2}{8(\pi/L)^3} = \frac{L^3 k^2}{2\pi^2} \qquad (2.12)$$

where we divided by 8 because we are only interested in positive wave numbers in the x, y and z directions. Therefore, bearing in mind that $k = \omega/c$, the density of modes (defined as the number of states divided by the volume) is given by

Density of Modes

$$\rho(\omega)d\omega \left(= \frac{n(k)}{V}dk\right) = \frac{\omega^2 d\omega}{2\pi^2 c^3} \qquad (2.13)$$

where $V = L^3$ is the volume of the cavity.[4] In reality this number will be twice as high since there are two possible polarization directions of light (horizontal and vertical — this is something that comes out of Maxwell's equations and the fact that the electric field is a vector, but is really a quantum mechanical property as we will see later on), so that finally we obtain

$$\rho(\omega)d\omega = \frac{\omega^2 d\omega}{\pi^2 c^3} \qquad (2.14)$$

Density Including Polarization

Notice that there is no restriction on the size of the frequency — it could in principle be arbitrarily large. Alternatively, the wave length of light inside a cavity can be arbitrarily small. There is, however, a limit on how big it could be $(2L)$.

2.2 Basic Properties of Lasers

First I will discuss basic properties of laser light that makes it very useful for us in optics as well as in applications to technology. Later on we will see how we can deduce these properties by using the quantum theory of light. Laser light has the following five important properties:

- *Directed.* The laser light is very focused and directed. This is because light inside a laser is an excellent approximation of a plane wave bouncing back and forth between two mirrors, and is then released gradually through a small hole. A laser of a diameter of a few centimeters directed at the surface of the Moon would only generate a spot of a size of a few hundred meters!

- *Intense.* Let's compare the intensity of an ordinary bulb with that of a laser. Laser intensities can easily reach $10^{10}W$, which is roughly 10^8 times higher than an ordinary bulb.

- *Monochromatic.* The light is very nearly of one color only. What does this mean exactly? This means that the frequency spread is only of the order of $10^6 Hz$, compared with the frequency of light produced, which is 10^{15}.

- *Coherent.* Coherence is a very important property if one is to study interference effects of light. There are two basic types of coherence which are important for laser light: spatial and temporal coherence. The first one has to do with two different

---

[4]I will use two different symbols for the density of states $\rho(\omega)$ and $\langle W_T \rangle$ throughout my notes (T stands for thermal). The meaning of the latter will become apparent in due course.

beams of light interfering at the same spatial location, while the second has to do with the same beam at one location, interfering with itself at two different times. We will talk about this in greater detail later in the book.

- *Short.* Laser can now generate pulses which are only a few femtoseconds in duration (10^{-15} seconds).

Why are lasers important? Because the light that comes out of them is very organized and structured. We need this kind of light if we are to manipulate matter with great precision. But what is going on inside a laser cavity? Inside an empty cavity, the state of radiation and matter is as chaotic as it can be[5] — blackbody radiation. Atoms move about in the lasing medium and interact with radiation which is bouncing back and forth between the walls. The radiation inside has many different frequencies (infinitely many) and many different phases. It is incoherent. It is also isotropic — the same in all directions. How can this be? We first devote ourselves to studying the disorder that gives rise to such a great order in the laser output.

Final Thought: I would like to say a few words about the history of lasers. In 1954 Townes in the USA and, independently, Basov and Prokorov in Russia, suggested a practical method of achieving lasing. This was using ammonia gas and produced amplified microwave radiation instead of visible light (called a maser). For this they shared the 1964 Nobel Prize for Physics. In 1958 Townes and Schawlow calculated the conditions to produce visible laser light. Finally in 1960 the first true laser was demonstrated by Maiman, using a ruby crystal. Since then, the usage of lasers has proliferated and this has greatly benefited and enhanced our understanding of the quantum nature of light. But before we can fly, we need first to learn how to walk.

---

[5]I will quantify this statement shortly. Briefly, by maximally chaotic, I will mean that the entropy is maximal under some constraints.

Chapter 3

Properties of Light: Blackbody Radiation

There are two basic ways in which heat propagates in a given medium: conduction and radiation.

Conduction is a relatively simple process. It is governed by a diffusion equation and the rate of change of temperature is proportional to the temperature gradient

Heat Conduction

$$\frac{dT}{dt} = -k\nabla^2 T \tag{3.1}$$

Once the temperature is the same everywhere there is no conduction — which is why I said that it was basically a relatively simple process.

Radiation, on the other hand, is independent of any temperature gradient and radiation processes are much more complicated to study. Even in equilibrium the behavior of light is not that simple, so much so that the intense study of its properties led to the advent of quantum mechanics!

Physicists like simple models and extreme situations. Limits of various types abound in physics, and as far as radiation is concerned, there are two useful situations — whitebody and blackbody radiation.

Whitebody radiation is when we have a body that reflects all the radiation that falls on it. Realistic bodies will of course be very reflecting only in some range of radiation and will be absorbing for other wavelengths. A mirror is a good example of a whitebody for visible radiation. Blackbody radiation is the other extreme. A blackbody absorbs all the radiation that falls on it. Surprisingly, there are a lot of good examples of near blackbodies, the Sun and the Earth to name a couple.

Suppose that we look at the radiation that leaves the blackbody. What kind of properties would it have. What would we see if we were inside a blackbody?

3.1 Planck's Quantum Derivation

Imagine that you are sitting in your room at night, reading by the night lamp. Your room is not in thermal equilibrium. Why? First of all, there is a lamp on. Switch the lamp off. Even then, there are objects in your room generating heat and emitting radiation that is not necessarily visible to you — yourself, for example. You are not in a thermal equilibrium — you generate heat, you are alive. But, suppose that you get rid of all sources and sinks of heat. What properties would the remaining radiation have that is in thermal equilibrium with the matter inside your room?

From Maxwell's equations we know that light is a wave: light is really — very loosely speaking at this stage — a collection of harmonic oscillators (to be proven rigorously later). In fact, at any temperature there are really infinitely many harmonic oscillators in any enclosed space, one for each possible frequency and polarization. So, your room as well as your oven is an infinite collection of "vibrating strings" each at a different frequency. What is the energy per oscillator in thermal equilibrium? Classical statistical physics tells us that every independent quadratic degree of freedom gets a $kT/2$ of energy[1] (the equipartition theorem — see any statistical mechanics textbook, e.g. [Huang (1963)]). The energy of a simple (one-dimensional) harmonic oscillator of frequency ω is given by

$$E_{HO} = 1/2(p^2 + \omega^2 x^2) \tag{3.5}$$

where ω is the (angular) frequency of the oscillator[2] and x and p are its position and momentum respectively. The energy density

---

[1]Here is why we get $kT/2$ per quadratic degree of freedom. Suppose that $E = a\xi^2/2$. The average energy is

$$\bar{E} = \frac{\int Ee^{-\beta E}dpdx}{\int e^{-\beta E}dpdx} = \frac{\int Ee^{-\beta E}d\xi}{\int e^{-\beta E}d\xi} \tag{3.2}$$

where $\beta = 1/KT$. Now, we can express this as

$$\bar{E} = -\frac{d}{d\beta}\log Z \tag{3.3}$$

where $Z = \int e^{-\beta E}d\xi$ is the partition function. We can then integrate this to obtain $Z = \beta^{-1/2}A$, where A is a constant. Hence $\log Z = const. - (\log\beta)/2$, so that

$$\bar{E} = \frac{1}{2\beta} = \frac{kT}{2} \tag{3.4}$$

[2]We will use interchangeably the angular frequency ω and the frequency $\nu = \omega/2\pi$.

is given by the famous (and ultimately wrong!) formula due to
Rayleigh and Jeans:

**Rayleigh–
Jeans
Formula**

$$u(\omega) = \rho(\omega)kT = \frac{\omega^2}{\pi^2 c^3}kT \qquad (3.6)$$

However, this formula implies that the total intensity becomes
larger and larger the higher the frequency, growing at the rate
proportional to the square of the frequency. In fact, the total intensity is infinite as the integral of the above quantity diverges.
This is clearly in contradiction with our everyday experience: just
open your oven and feel that the heat coming out of it is actually
finite (also, neither the Earth nor the Sun radiate infinite amounts
of energy or else we wouldn't be here to talk about it). Therefore,
according to classical physics, we reach a paradoxical conclusion.

In order to correct this very embarrassing mistake, Planck pos-
tulated the following weird assumption: the harmonic oscillator
can have energies only in the "packets" of $\hbar\omega$ — no continuum of
energies is allowed (this in fact replaces the integral in footnote 1
by a discrete sum and eventually gets rid of infinities as we will see
shortly).

Planck's derivation provides a real insight into the way physicists think, which is why I will cover it in a bit more detail. He
knew from experiments the exact shape of the curve for the blackbody radiation density and from that he had extracted a formula
which fitted the curve very well and is plotted in Figure 3.1.

The formula which seemed to fit the curve of blackbody radiation density was something like

$$u(\omega) \propto \frac{\omega^3}{e^{\hbar\omega/kT} - 1} \qquad (3.7)$$

Notice that this formula doesn't have the feature that it "blows
up" for large frequencies — instead it approaches zero as $\omega \to \infty$.
Planck then saw that a way of deriving this formula was to assume that the energies of the harmonic oscillator are quantized. He
needed another assumption to complete his derivation. He needed
to assume that the probability to occupy the level with energy E
is

$$P(E) = e^{-E/kT}/Z \qquad (3.8)$$

where $Z = \sum_E e^{-E/kT}$ is the partition function, which was known
to him from Boltzmann's (completely classical, of course) work.

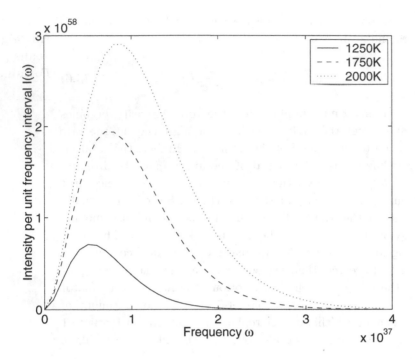

Fig. 3.1   Different blackbody spectra at different temperatures. The area under each of the curves has a finite value unlike that predicted by classical physics (statistical mechanics).

Then, the total (i.e. average) energy is given by (where N_0 is the total number of oscillators)

$$\langle E \rangle = N_0 \frac{\sum_n \hbar\omega n e^{-n\hbar\omega/kT}}{Z} \tag{3.9}$$

$$= \frac{N_0\hbar\omega(0 + q + 2q^2 + 3q^3 + \cdots)}{N_0(1 + q + q^2 + \cdots)} \tag{3.10}$$

where

$$q = e^{-\hbar\omega/kT} \tag{3.11}$$

Therefore, the average energy per oscillator (i.e. divided through by N_0) is

$$\langle E \rangle = \frac{\hbar\omega}{e^{\hbar\omega/kT} - 1} \tag{3.12}$$

(A way of obtaining the final expression is to start with the well-known sum for the geometric series

$$\sum e^{-\alpha n} = \frac{1}{1 - e^{-\alpha}} \tag{3.13}$$

and then differentiate both sides by $d\alpha$ to obtain

$$\sum ne^{-\alpha n} = \frac{e^{-\alpha}}{(1 - e^{-\alpha})^2} \qquad (3.14)$$

This is all we need to derive the average energy.)

At high temperatures the average energy per oscillator becomes the expected kT of energy per oscillator in accordance with classical physics (prove it). So, this is the correct average energy we should be using for quantum bosonic systems (bosonic systems are ones consisting of particles with integral spin).

The intensity per unit frequency is now found to be **Blackbody Spectrum**

$$\begin{aligned} I(\omega)d\omega &= cu(\omega)d\omega \\ &= c\langle E \rangle \rho(\omega)d\omega \\ &= \frac{\hbar\omega^3 d\omega}{\pi^2 c^2 (e^{\hbar\omega/kT} - 1)} \end{aligned} \qquad (3.15)$$

Note that, as we have said before, there is nothing strange happening at large frequencies. I on its own is the intensity per unit frequency interval. I can be multiplied by some frequency interval to obtain the intensity. In fact, the derivation just presented is really due to Ehrenfest (and refined by Einstein) not Planck. (Einstein had improved Planck's derivation considerably and made it much easier to understand and manipulate). But the bottom line is, of course, the same: *light is quantized*. However, in the first part of the book we won't really have to use this assumption extensively! The reason is that a great deal of results can be derived without this assumption.

Note briefly that the total output intensity from a blackbody is obtained by integrating Planck's expression

$$I = \int_0^\infty I(\omega)d\omega = \frac{(kT)^4}{\hbar^3 \pi^2 c^2} \int_0^\infty \frac{x^3 dx}{e^x - 1} \qquad (3.16)$$

The value of the integral is $\pi^4/15$ so that **Stefan–Boltzmann Law**

$$I = \frac{\pi^2}{15\hbar^3 c^2}(kT)^4 \qquad (3.17)$$

which is the well-known Stefan–Boltzmann law $I = \sigma T^4$, where $\sigma = \pi^2 k^4/15\hbar^3 c^2$ is Stefan's constant. So the total intensity is finite, unlike in classical physics, and fits perfectly all the experimental evidence so far. Many systems radiate according to this formula, and more recently we have seen pictures of the cosmic background radiation following this formula with the temperature of $2.7K$. So, the universe is an excellent blackbody.

I'd like to discuss the blackbody spectrum derivation a bit more. We see that the blackbody radiation is produced by a bunch of harmonic oscillators. But what is it that oscillates? Radiation is an electromagnetic wave and it therefore oscillates. But recall the way we phrased the equipartition: the energy is that of an atom, its position x and its momentum p. It turns out (and we will do this in much more detail) that atoms (in the walls of the cavity) also oscillate. As far as the blackbody radiation is concerned, they are also harmonic oscillators, and this is in general a remarkably accurate picture (to be seen later). The radiation and atoms are all harmonic oscillators in equilibrium and this is what produces the right spectrum as described above. This is also what was used later on by Einstein to give the first (almost) quantum model for atom–light interaction (Chapter 4). Before I go into that, I'd like to present the modern way of viewing Planck's result, leading to the Bose–Einstein (and Fermi–Dirac) statistics.[3]

3.2 The Proper Derivation of Planck's Formula

We will now derive probabilities of occupying various energy levels when the particles are obeying the laws of quantum mechanics, but now by using a highly sophisticated method of statistical mechanics. This process is based on maximizing entropy under certain given constraints.[4] I should say in passing that the majority of professional physicists do not realize that this method is really the only sound and rigorous mathematical way of handling probabilities in statistical mechanics and that it has a deep mathematical justification in Bayesian inference method.[5] We know, of course, that all the particles are *always* really quantum mechanical, but we will see that at high temperature their quantum properties are not manifested and the statistical results coincide with Gibbs' probabilities (i.e. they follow the $e^{-E/kT}$ law). *It is very interesting to note that the difference between the quantum and classical statistics will only be in the manner we count the number of different ways that particles can occupy various different states.* I will describe both

---

[3]You may have seen this in a statistical mechanics book.

[4]When there are no constraints the resulting probabilities form the so-called microcanonical ensemble, when the total energy is fixed we have the canonical ensemble, and finally when both the energy and particle numbers are fixed we have the grand canonical ensemble.

[5]In fact, anyone betting money on the stock market will follow this theory, which is the best indication that it has safe foundations!

bosons and fermions for completeness so that we can understand the difference more clearly. First we start with fermions, and then talk about how bosons (e.g. light) behave in much more detail.

Suppose that we have energy levels $E_1, E_2, \ldots$, with degeneracies $g_1, g_2, \ldots$.[6] We have N fermions which must be distributed among these levels such that the total energy is E. The occupation numbers are denoted by $n_1, n_2, \ldots$ so that we have the following restrictions (this is the so-called grand canonical ensemble that you can read about in any statistical mechanics textbook):

$$\sum_k n_k = N \tag{3.18}$$

$$\sum_k n_k E_k = E \tag{3.19}$$

Since the particles are fermions they obey the *Pauli exclusion principle* which implies that $n_k \leq g_k$. In addition, quantum particles are completely *indistinguishable*. We can never tell one particle from another. Not even Nature knows which is which: quantum systems suffer from a complete identity crisis. We thus consider the problem of distributing n_k indistinguishable particles into g_k *distinguishable* states. The number of different ways of distributing n_k fermions into g_k states is (make sure you understand this and can reproduce it)

$$\frac{g_k!}{n_k!\,(g_k - n_k)!} \tag{3.20}$$

The total number of ways of distribution of N Fermi particles among all of the energy levels is therefore given by

$$\Omega_F = \prod_i \frac{g_i!}{n_i!(g_i - n_i)!} \tag{3.21}$$

This is the quantity that we will be manipulating to obtain various occupation probabilities.

Let us now perform the same calculation for bosons. We now have the same situation as above apart from the fact that there is no restriction on how many particles can occupy one state (no Pauli exclusion principle exists here). The total number of ways of distributing N bosons into g_k states given that more that one boson can occupy the same state is given by (again, make sure you

---

[6]Here we will perform the most general calculation assuming that there exist many different levels with the same energy. This assumption will always be omitted later on in the book and we will always assume that $g = 1$.

can reproduce and justify this expression)

$$\Omega_B = \prod_i \frac{(g_i + n_i - 1)!}{n_i!(g_i - 1)!} \tag{3.22}$$

The idea is now to maximize entropy subject to the energy and particle number constraints (this is as I said the only sound way of obtaining probabilities when there is uncertainty in the overall state of your system). Using the method of Lagrange multipliers,

$$\frac{\partial}{\partial n_k}\left(\ln \Omega_{F,B} - \alpha N - \beta E\right) = 0 \tag{3.23}$$

This leads to the following (prove it!)

$$\frac{n_k}{g_k} = \frac{1}{e^\alpha e^{\beta E_k} + 1} \quad \text{(fermions)} \tag{3.24}$$

$$\frac{n_k}{g_k} = \frac{1}{e^\alpha e^{\beta E_k} - 1} \quad \text{(bosons)} \tag{3.25}$$

It turns out that a lengthy manipulation of these expression together with the given constraints yields the final result (prove it!)

$$\frac{n_k}{g_k} = \frac{1}{e^{(E_k - \mu)/kT} + 1} \quad \text{(fermions)} \tag{3.26}$$

$$\frac{n_k}{g_k} = \frac{1}{e^{(E_k - \mu)/kT} - 1} \quad \text{(bosons)} \tag{3.27}$$

In the case of photons, their number is not conserved, which means that the chemical potential $\mu = 0$ (this is because the second constraint doesn't really apply to photons). When there is no degeneracy, $g_k = 1$, we recover the well-known formula

$$n_k = \frac{1}{e^{E_k/kT} - 1} \tag{3.28}$$

Bose–Einstein Distribution

This formula will allow us to understand more properties of radiation now, and, in particular, fluctuations in the number of photons. This quantity, as discovered by Einstein, somehow contains the central mystery of quantum mechanics — it combines both the wave and particle nature of light into one.

3.3   Fluctuations of Light

From Planck's distribution we know that the average number of photons in the blackbody radiation is given by

$$\langle n \rangle = \frac{\langle E \rangle}{\hbar \omega} = \frac{1}{e^{\hbar \omega/kT} - 1} \tag{3.29}$$

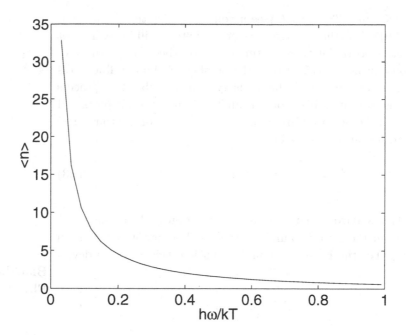

Fig. 3.2 Average photon number for blackbody radiation as a function of frequency divided by temperature.

The average number of photons for blackbody radiation is plotted in Figure 3.2. Let's try to get a feel for the size of the numbers involved by considering two of the most familiar sources for us. First, a conventional incandescent lamp is at $T = 3000K$, and so $kT \approx 0.3keV$. For optical frequencies we have $1.5eV \leq \hbar\omega \leq 2.5eV$ and so $\langle n \rangle = 10^{-3}$. For the Sun, on the other hand, the surface temperature is $T = 5000K$, so that $\langle n \rangle = 10^{-2}$.

From equation (3.29) we can infer the probability p_n of having n photons since we know from elementary statistics that the average number of photons is given by

$$\langle n \rangle = \sum_n n p_n \tag{3.30}$$

and so we obtain

$$p_n = \frac{\langle n \rangle^n}{(\langle n \rangle + 1)^{n+1}} \tag{3.31}$$

Bose–Einstein Distribution (Expression II)

because

$$p_n = \frac{e^{-\beta n}}{\sum_n e^{-\beta n}} = (1 - e^{-\beta})e^{-\beta n} \tag{3.32}$$

where, as usual, $\beta = \hbar\omega/kT$ (see if you can derive this from equation (3.29)). But the average number of photons will in reality not be constant as radiation is continuously absorbed and re-emitted. The consequence of this is that the number of photons fluctuates about the average value. The quantity that describes this process is another very important one in understanding the properties of light. It is the deviation from this mean number of photons. The standard deviation σ is defined by

$$\sigma^2 := \sum_n p_n(n - \langle n \rangle)^2 = \langle n^2 \rangle - \langle n \rangle^2 \qquad (3.33)$$

Thus the standard deviation tells us how far away the actual number is from the average number (that is, how far it deviates on average). For the Bose–Einstein distribution, the standard deviation is

Blackbody Fluctuations

$$\sigma^2 = \langle n \rangle + \langle n \rangle^2 \qquad (3.34)$$

(Prove this result using the fact that the average number of photons can be expressed in two different ways as

$$(\langle n \rangle =) \frac{1}{e^\beta - 1} = \sum_n ne^{-\beta n}(1 - e^{-\beta})$$

and differentiating both sides of this expression by $d/d\beta$. One of the terms on the right hand side will then be $\langle n^2 \rangle$, and this is exactly what is needed for the standard deviation.)

Equation (3.34) is an interesting formula. How can we interpret it? Let's first ask ourselves what kind of standard deviation we would expect from particles in a box. Imagine that we have a box partitioned into N compartments such that there are n particles distributed there. The probability that a particular particle is situated in the kth compartment is $1/N$ independently of k. The average number of particles per compartment is n/N. The number of possible ways of choosing j particles out of n into one compartment is $\binom{n}{j}$. Putting these statements together, the probability of having j particles in the kth compartment is

$$p_j = \binom{n}{j}\left(\frac{1}{N}\right)^j\left(1 - \frac{1}{N}\right)^{n-j} \qquad (3.35)$$

Let's now calculate the standard deviation in the number of

particles. Introduce the following notation $\omega = 1/N$. Then,

$$\langle n \rangle = \sum_j j p_j = \sum_j \frac{jn!}{j!(n-j)!} \omega^j (1-\omega)^{n-j} \qquad (3.36)$$

$$= n\omega \sum_{j=1}^{n} \frac{(n-1)!}{(j-1)!(n-1)-(j-1)!} \omega^{j-1} (1-\omega)^{(n-1)-(j-1)}$$

Invoking the well-known binomial theorem this reduces to

$$\langle n \rangle = n\omega \qquad (3.37)$$

as we might expect from our knowledge of secondary school statistics. Next we need to calculate $\langle n^2 \rangle$. We notice that it is easier to calculate

$$\langle n(n-1) \rangle = \sum_j j(j-1) p_j$$

$$= n(n-1)\omega^2 \sum_{j=2}^{n} \frac{(n-2)!}{(j-2)!(n-2)-(j-2)!}$$

$$\times \omega^{j-2} (1-\omega)^{(n-2)-(j-2)} \qquad (3.38)$$

Therefore,

$$\langle n(n-1) \rangle = n(n-1)\omega^2 \qquad (3.39)$$

So now we can use this result to calculate what we need

$$\langle n^2 \rangle = \langle n(n-1) \rangle + \langle n \rangle = n(n-1)\omega^2 + n\omega \qquad (3.40)$$

The standard deviation can therefore be calculated to be

Particle Fluctuations

$$\sigma^2 - n\omega(1-\omega) \qquad (3.41)$$

For large N σ^2 tends to

$$\frac{n}{N} = \langle n \rangle \qquad (3.42)$$

which is the average number of particles per partition. Thus the spread of the number of particles around the mean is the size of the root of the mean. This is a well-known result for particles.

In order to perhaps make this result a bit more transparent, let's imagine that the box is partitioned into two — left and right — compartments. The standard deviation tells us about the average difference between the left and right parts. Now imagine that we are randomly deciding to place each particle left or right. Then after i steps the difference is given by

$$\Delta_i = N_i^L - N_i^R \qquad (3.43)$$

where N_i^L and N_i^R are the numbers of particles on the left and right respectively after i steps. We also know that the difference in the previous step could have been either by one particle larger or smaller than this, so that

$$\Delta_i = \Delta_{i-1} \pm 1 \qquad (3.44)$$

Squaring both sides of this equation we obtain

$$\Delta_i^2 = \Delta_{i-1}^2 \pm 2\Delta_{i-1} + 1 \qquad (3.45)$$

Taking the average of both sides we obtain

$$\langle \Delta_i^2 \rangle = \langle \Delta_{i-1}^2 \rangle + 1 \qquad (3.46)$$

since the average of the difference Δ is zero (the process is random). As $\sigma_i^2 = \langle \Delta_i^2 \rangle$, we obtain the following recursive relationship for the standard deviation:

$$\sigma_i^2 = \sigma_{i-1}^2 + 1 \qquad (3.47)$$

We can therefore conclude that

$$\sigma_n^2 = n \qquad (3.48)$$

and so $\sigma_n = \sqrt{n}$ as before. So, we confirm again that the "square root" behavior in the deviation is typical for particles.

How about waves in the box? Well, here we have a different scenario. Suppose that we have n different waves in a cavity. In the state of thermodynamic equilibrium, the waves will all have different frequencies and thus phases:

$$E_k = E_0 e^{i\phi_k} \qquad (3.49)$$

The total field is

$$E = \sum_i E_i \qquad (3.50)$$

and the intensity (proportional to the number of photons) is

$$I = E^*E = \sum_{jk} |E_0|^2 e^{-i\phi_j} e^{i\phi_k} \qquad (3.51)$$

$$= |E_0|^2 \left(n + \sum_{j \neq k} e^{i(\phi_k - \phi_j)} \right) \qquad (3.52)$$

$$= |E_0|^2 \left(n + 2 \sum_{j > k} \cos(\phi_k - \phi_j) \right) \qquad (3.53)$$

The average of cos is zero since phases are random, so that

$$\langle I \rangle = |E_0|^2 n \qquad (3.54)$$

To compute the standard deviation we need the average of

$$(|E|^2)^2 = |E_0|^4 \left(n + \sum_{j \neq k} e^{i(\phi_k - \phi_j)} \right)^2 \qquad (3.55)$$

which on average leads to

$$\langle (|E|^2)^2 \rangle = |E_0|^4 (n^2 + n(n-1)) \approx 2|E_0|^4 n^2 \qquad (3.56)$$

(remember that we are considering a large number of waves which is why both the averaging and the second equality in the above equation are justified). Therefore the standard deviation is given by

$$\sigma^2 = |E_0|^4 n^2 \qquad (3.57)$$

Finally, the standard deviation in number is **Wave Fluctuations**

$$\sigma^2 \sim n^2 \qquad (3.58)$$

where n is the total number of waves. Going back to the standard deviation in the blackbody radiation we see that for small average number of photons (low temperature, high frequency), it behaves as if it is made up of particles (the quantum regime) while for large n it behaves more like a wave (the classical regime). Remarkably, the right behavior for any regime is given by the sum of the two extrema — waves and particles at the same time! Behavior of the fluctuations of light therefore goes right to the heart of the quantum matter and the notion of the "wave particle duality". Marvellous!

Final Thought: Making breakthroughs in science requires a great deal of effort and guessing. The process leading up to the final triumph is not as clean as usually presented in textbooks. Planck had to display a great deal of courage to challenge the orthodox view of continuous energy emission and absorption presented by classical physics. But none of this, although quite necessary, is in fact sufficient. What is important in the end is whether the result is confirmed by experiment or not. Planck's quantum hypothesis has never been contradicted in the slightest in the last 100 years since its birth.

So, light is in some sense a particle and a wave. And, if one of these two aspects were missing in our theoretical description, this would be contradicted by our experiments. However — and this is very interesting and happens often in physics — other theories in physics would also be contradicted if light was not a wave and particle at the same time. I will show now that if light was only a particle, we could then be able to construct the perpetual machine of the second kind (i.e. we could violate the second law of thermodynamics).

3.4   Maxwell's Lucifer

We will now use observation by scattering light to construct a perpetuum mobile of the second kind, i.e. a machine that operates in a cycle and violates the second law of thermodynamics! This is, of course, impossible (or, more precisely, it is highly unlikely) as the second law of thermodynamics is one of the most experimentally tested and trusted laws of physics. But it is very hard to see where the flaw is which is one of the reasons I talk about it. The other reason is that it will turn out that if light was indeed made up of classical particles, this perpetuum mobile would work. It is the wave nature of light that prevents us from using energy in a lossless way!

Have a look at the machine presented in Figure 3.3. This smart proposal is due to Gabor[7] and, remarkably, it is related to the operation of a laser.[8]

We proceed with the description of the Lucifer imagined by Gabor. Most proposals for violating the second law consist of two parts: the first one is making an observation on the system used for violation and the second part involves acting on the system based on this information in order to obtain as much work out of it as

---

[7]who, incidentally, received a Nobel Prize in 1971 for inventing holography.

[8]The title of the section comes from the fact that Maxwell was the first person to worry about a small and smart being which could violate the second law by being able to record all the positions and velocities of molecules he was using in his engine. This being has subsequently been named Maxwell's demon, and since our particular demon will be using light for his observation, we will call it Maxwell's Lucifer!

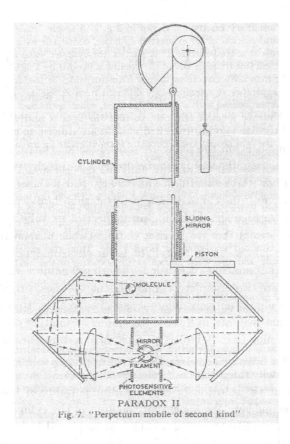

CYLINDER

SLIDING MIRROR

PISTON

"MOLECULE"

MIRROR

FILAMENT

PHOTOSENSITIVE ELEMENTS

PARADOX II

Fig. 7. "Perpetuum mobile of second kind"

Fig. 3.3 Maxwell's Lucifer designed by Gabor to violate the second law of thermodynamics. The workings of this machinery are explained in the text.

possible. In Gabor's case we have a molecule which we use to push a piston, thereby doing work. For the atom to push the piston in the right way we need to determine its location accurately. This process consists of two key stages:

- making sure that the molecule is in the volume V_1 (in Figure 3.3, this is the volume underneath the piston which the molecule occupies);
- sliding the partition to make sure that the molecule does some work in pushing the piston up.

The second law is violated if the amount of heat dissipated in observing the position of the molecule is less than the amount of work gained by that molecule doing the work in pushing the piston up. Thermodynamics tells us that in a reversible adiabatic expansion the entropy reduction is given by

$$\Delta S = k \ln \left\{ \frac{V_1}{V} \right\} = -k \ln \left| \frac{V_1}{V} \right| \tag{3.59}$$

and its negative value signifies that entropy is reduced (V is the total volume of the cylinder in Figure 3.3). It seems that the amount of energy lost and the corresponding entropy increase due to observation by scattering light off the molecule can be made arbitrarily small and hence the overall entropy can be decreased, thereby violating the second law (the second law equivalently states that the entropy cannot decrease in a closed system). But this should not be possible, or at least we should be highly suspect of its validity. There must be a catch. It turns out that the fundamental reason why the Lucifer is impossible is that light is an electromagnetic wave! But this is an amazing conclusion. Thermodynamics and electromagnetism are two completely independent theories, originally developed to address entirely different aspects of reality. Here, however, we see that they are in a sense dependent. Violating one of them (i.e. assuming that light has just particle nature), leads to violation of the other one (i.e. decrease in entropy of a closed system). Let's see why this is so.

We now present the solution of the apparent paradox of the violation of the second law. We know that the light in the cylinder is a wave and behaves like waves in a cavity we analyzed before. The wavelengths that can exist in the cylinder have to be *discretized* according to the usual formula

$$\lambda = 2L, L, \frac{2L}{3}, \ldots, \frac{2L}{M} \tag{3.60}$$

where L is the length of the volume of the cavity between the four mirrors where the light used for observation is trapped. In order to be able to determine precisely if the molecule is in the volume V_1 we have to make sure that the wavelength should not be shorter than $2L_1$, this being the length of the volume V_1. The reason for this is that if the wavelength is smaller, the momentum kick transferred to the molecule due to interaction with light will be bigger and we cannot be certain that the molecule is still in V_1. This implies that

$$M = \frac{L}{L_1} = \frac{V}{V_1} \tag{3.61}$$

is the total number of modes in this light packet. How will the observation be performed? We can send pulses of length τ separated in time by t_u, as shown in Figure 3.4.

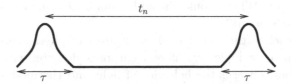

Fig. 3.4   Pulses of light of length τ separated by time t_u.

In order to make sure that we have a chance of observing the molecule we should have

$$\frac{t_u}{\tau} = \frac{V}{V_1} \tag{3.62}$$

i.e. the frequency of light pulses should be at least as big as the frequency of molecule's presence in the volume V_1. Now we are in a position to calculate the entropy increase due to this observation. Its value per mode is given by

$$\frac{\Delta S}{M} = \frac{\Delta Q}{T} = \frac{\frac{1}{2}kT}{T}\frac{t_u}{\tau} \tag{3.63}$$

where every pulse gets $1/2kT$ and there are t_u/τ pulses. This formula is based on the standard thermodynamical expression $dQ = TdS$. Therefore the total loss in all M modes is

$$\Delta S = \frac{Mkt_u}{2\tau} = \frac{k}{2}\left(\frac{V}{V_1}\right)^2 \tag{3.64}$$

Entropy Increase Due to Observation

Using the well-known inequality $x^2 \geq \ln x^2$ we arrive at

$$\frac{k}{2}\left(\frac{V}{V_1}\right)^2 \geq \frac{k}{2}\ln\left(\frac{V}{V_1}\right)^2 \tag{3.65}$$

which shows that the entropy increase due to scattering of light is greater than the entropy decrease in the cycle. Hence the second law is saved! Maxwell's Lucifer is therefore unable to act in this way to violate thermodynamical laws, and the basic reason is because light is an electromagnetic wave. Note that quantum mechanics does not help either — and Gabor knew this. The reason is that every photon behaves in this wavelike way. Also, it makes no difference if some other particles were used instead of photons — they all behave in the same way according to the rules of quantum mechanics. An interesting open question is what would happen if we tried to use the interaction-free measurement I described before to observe the molecule. Can you see how the second law would be "saved" in this case? Where would the additional heat come from to keep the books straight?

I now come back to the fact that different areas of physics, although originating for completely different purposes, do agree with each other. Optics concerns the behavior of light and thermodynamics was invented to describe the behavior of steam engines and their efficiency. At first sight these two areas have very little, if anything, in common. Yet, we saw that if we try to change the laws of optics, this will automatically affect the laws of thermodynamics. This may reaffirm us in our belief that all of physics describes one and the same world (dare I say "reality"), albeit different aspects of it. It is also what makes physics difficult to modify, as it means that we cannot change the rules arbitrarily in one part, and trust that the rest is unaffected.

Chapter 4

Interaction of Light with Matter I

Einstein thought about the interaction of light and matter in 1917, well before the advent of proper quantum mechanics in 1925. So, let's put ourselves in his position and see what kind of information was available to him to attack this problem. First of all, he knew about Planck's quantum assumption and the correct derivation of a blackbody's spectrum. Secondly, he knew that atoms were also quantized and that electrons occupied stationary states as in Bohr's "planetary" atomic model. Fine. But he had no clue about how atoms interacted with photons (neither do we at this stage in the book). He only knew that an atom can absorb a photon and move from a lower energy level to a higher energy level providing that the frequency of the photon is equal to the energy difference between the levels — this process he called *stimulated absorption*. He also knew that an atom can spontaneously emit a photon and jump from a higher energy level to a lower one — the process called *spontaneous emission*. So, then he reckoned that at thermal equilibrium the rate of emission and absorption have to be equalized.

Assume first that we are looking at two levels only (this simplest model considered by Einstein contains all the features of the most general model). We will label the levels as 1 and 2^1 as shown in Figure 4.1. Assume that we only have stimulated absorption from the ground state 1 and spontaneous emission from the excited state 2. This is a very natural assumption. If we supply energy (photons) to this atom, then it will absorb it and jump from 1 to 2. It was known to Einstein that if an atom is left for long enough in an excited state it will naturally (spontaneously) emit energy and relax to the ground state. The rate of emission is given by A_{21} (known as Einstein's A coefficient), so that the total number

---

[1] We will also refer to these as the ground state and the excited state.

of atoms spontaneously emitting per unit of time is $A_{21}N_2$, where N_2 is the number of atoms in level 2. The rate of absorption is, on the other hand, also proportional to the density of radiation $u(\omega_{12})$,[2] so that the total number of atoms absorbed per unit of time is $u(\omega_{12})B_{12}N_1$, where B_{12} is the Einstein B coefficient and N_1 is the number of atoms in the level 1. In equilibrium the two rates are equal (*by definition*), so that

$$u(\omega_{12})B_{12}N_1 = A_{21}N_2 \qquad (4.1)$$

Therefore,

$$u(\omega_{12}) = \frac{A_{21}N_2}{B_{12}N_1} \qquad (4.2)$$

But, according to Boltzmann we have that

$$\frac{N_2}{N_1} = e^{-(E_2-E_1)/kT} = e^{-\hbar\omega_{12}/kT} \qquad (4.3)$$

Thus, following this logic Einstein would have obtained

Wrong Energy Density Formula

$$u(\omega_{12}) = \frac{A_{21}}{B_{12}}e^{-\hbar\omega_{12}/kT} \qquad (4.4)$$

But, this is obviously *wrong* as the right formula for the radiation density should read (according to Planck)

$$u(\omega_{12}) = \frac{\hbar\omega_{12}^3}{\pi^2 c^3}\frac{1}{e^{\hbar\omega_{12}/kT} - 1} \qquad (4.5)$$

You can simply imagine that this two level atom is sitting in the wall of the blackbody cavity and therefore it has to produce the same kind of radiation that exists inside, since the walls and the radiation are in equilibrium. The two expressions in equation (4.4) and equation (4.5) coincide only at small temperature or high frequency (prove it). From this correspondence at small T, we can conclude that

$$\frac{A_{21}}{B_{12}} = \frac{\hbar\omega_{12}^3}{\pi^2 c^3} \qquad (4.6)$$

So we can obtain the right relationship between spontaneous emission and stimulated absorption even though we have obviously made a mistake, i.e. we haven't been able to reproduce the right radiation formula of Planck's![3]

---

[2]Spontaneous emission is independent of radiation density — even if there is no radiation around the atom will still emit!

[3]And we don't have the right quantum theory either. So, two wrongs do make a right after all!

4.1 Stimulated and Spontaneous Emission

We now need to write a new detailed balance that will result in the correct blackbody formula. It is not enough to have stimulated absorption and spontaneous emission — we need another process to be able to satisfy Planck's formula as we saw in the last chapter. This is exactly what forced Einstein to add another process: stimulated emission. The three processes are shown in Figure 4.1.

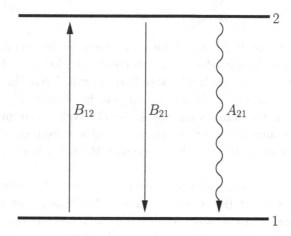

Fig. 4.1 Two level atom in Einstein's model. There are three processes present: stimulated emission and absorption and spontaneous emission. All three are necessary to predict the correct blackbody spectrum as explained in the notes.

The (now correct) detailed balance that Einstein wrote down in his paper was

$$\frac{dN_2}{dt} = -N_2 A_{21} - N_2 B_{21} u(\omega_{12}) \tag{4.7}$$

$$+ N_1 B_{12} u(\omega_{12}) \tag{4.8}$$

The Correct Rate Equation

where $N_2 B_{21} u(\omega_{12})$ is the number of atoms that emit *in a stimulated fashion* per unit of time (write down the analogous formula for dN_1/dt). As we have seen, Einstein needed this term in order to obtain the right density of the radiation field. Therefore, Einstein concluded that the process of stimulated emission must also exist. It is amazing that such an important conclusion can be derived from such a simple argument. This is a typical way that physicists think and we will encounter it again and again: first make a conjecture, then compute its consequences and see if they agree with the already known results (e.g. Planck's radiation formula);

if not, modify the conjecture until the result is correct! We note that stimulated emission is highly counterintuitive as by supplying energy to the atom we would expect it to get more (and not less) excited. But obviously, our classical intuition fails in this case. The equilibrium condition now implies that $dN_2/dt = 0$, so that

$$u(\omega_{12}) = \frac{A_{21}}{\frac{N_1}{N_2}B_{12} - B_{21}} \qquad (4.9)$$

$$= \frac{A_{21}}{B_{21}} \frac{1}{\frac{B_{12}}{B_{21}}(e^{\hbar\omega_{12}/kT} - 1)} \qquad (4.10)$$

It follows therefore that $B_{21} = B_{12}$. This is another amazing result from such a simple theory: the rate of stimulated emission and absorption are equal to each other! (Note that we will derive this rigorously using a quantum mechanical treatment in Chapter 7.)

But what about the actual values of A and B? Well, Einstein was unable to say anything about them: he lacked a more precise quantum mechanical formulation that came only 10 years after his calculation.

Let's now look at one particular extreme behavior in the rate equations. Suppose that the density of light is made larger and larger. The atoms would be able to follow this increase for some time by absorbing more and more, but they would ultimately reach their maximum capacity when all atoms become excited. At this point light would just continue to propagate through material without being absorbed by the atoms as they are saturated. The saturation radiative density, W_S, is defined so that the rate of spontaneous and stimulated emissions are equal. This means that

Saturation Density

$$W_S B = A \qquad (4.11)$$

which leads to the following formula[4]

$$W_S = \frac{\hbar\omega^3}{\pi^2 c^3} \qquad (4.12)$$

which is the density that equalizes the spontaneous and stimulated emission rates and is the same as the expression in equation (4.6).

A two level atom therefore either emits or absorbs light of energy which is equal to the energy difference between the two levels (strictly speaking this equality of photon's energy and the atomic

---

[4]I now use a different symbol for the radiation density as we are going to move onto a more general treatment in the next section. The symbol $u(\omega)$ will be used for the thermal (blackbody) radiation density, while the more general notation W will symbolize the density of radiation when other sources are present as well (lasers for example).

energy difference is not necessary, as we will see later). Total energy is in this case conserved. How about the momentum conservation? Well, in stimulated emission the light is emitted in the same direction as the absorbed light, so that the total net momentum transfer due to stimulated processes is zero. Suppose that the atom is initially stationary. This immediately tells us that since the momentum has to be conserved, spontaneously emitted light cannot be in any particular direction. Think about a gas of atoms and suppose that spontaneous emission was directed in a particular way. Then you would expect the gas to move (drift) in a particular direction (i.e. the center of mass would be moving). But this never happens in reality. Therefore spontaneous emission has to be random. A more detailed and careful quantum calculation shows that this is indeed the case, and spontaneous emission is completely random, i.e. uniformly distributed over the 4π solid angle centered on the atom. This fact will have important consequences in the process of *laser cooling* of atoms. Looking ahead, we know that light is made up of photons. What does this uniform distribution mean in this case? It means that if we surround the atom with light detectors, then which particular one will register a photon is completely random (but, of course, only one detector will click at any instance of time).

4.2   Optical Excitation of Two Level Atoms

Now imagine that our cavity contains a certain medium and in addition to the blackbody radiation present in equilibrium we have an external source of radiation interacting with the atoms in the medium. (Here I mainly follow the approach of Loudon in [Loudon (1964)]). We can imagine that this external source is another laser or some other source of heat or energy in general, like a lamp. Having an external field interacting with a two level atoms leads to the following rate equation

External Field Rate Equation

$$\frac{dN_1}{dt} = -\frac{dN_2}{dt} = N_2 A + (N_2 - N_1)B\langle W \rangle \qquad (4.13)$$

where now the radiation density $\langle W \rangle$ has two independent components

$$\langle W \rangle = \langle W_T \rangle + \langle W_E \rangle \qquad (4.14)$$

the thermal blackbody density $\langle W_T \rangle$[5] and the external density

_____

[5]This is what we called $u(\omega)$ previously, but now I change the notation as

$\langle W_E \rangle$ (the latter did not exist in Einstein's treatment). At room temperature, the contribution from the blackbody radiation will be in general much smaller than from an external source such as another laser. Therefore we assume that the blackbody radiation is absent. Suppose that all the population is initially in the ground state (i.e. state 1). Then solving the rate equation (4.13) we obtain (prove it)

$$N_2(t) = \frac{NB\langle W \rangle}{A + 2B\langle W \rangle}(1 - e^{-(A+2B\langle W \rangle)t}) \qquad (4.15)$$

For short times we obtain (prove it)

$$N_2 = NB\langle W \rangle t \qquad (4.16)$$

i.e. the number of excited atoms increases linearly with time. For times long enough, the number of atoms approaches its steady state value. Let's see what happens in a steady state case. Then, by definition from equation (4.13), we have   **Steady State**

$$N_2 A + (N_2 - N_1)B\langle W \rangle = 0 \qquad (4.17)$$

We now use the fact that $W_S B = A$, so that

$$N_2 W_S + (N_2 - N_1)\langle W \rangle = 0 \qquad (4.18)$$

Therefore as $N = N_1 + N_2$ we have

$$N_1 = \frac{A + B\langle W \rangle}{A + 2B\langle W \rangle}N = \frac{W_S + \langle W \rangle}{W_S + 2\langle W \rangle}N \qquad (4.19)$$

$$N_2 = \frac{B\langle W \rangle}{A + 2B\langle W \rangle}N = \frac{\langle W \rangle}{W_S + 2\langle W \rangle}N \qquad (4.20)$$

This leads to the following steady state rates for emission and absorption

$$N_1 B\langle W \rangle = \frac{(W_S + \langle W \rangle)\langle W \rangle}{(W_S + 2\langle W \rangle)W_S}NA \qquad (4.21)$$

$$N_2 B\langle W \rangle = \frac{\langle W \rangle^2}{(W_S + 2\langle W \rangle)W_S}NA \qquad (4.22)$$

$$N_2 A = \frac{\langle W \rangle}{W_S + 2\langle W \rangle}NA \qquad (4.23)$$

When we turn on the laser initially, the population in the excited state starts to increase linearly and it finally reaches its steady state

we give it a more general treatment.

value. In the steady state the populations don't change any more and the total amount of energy stored in the atoms is given by

$$N_2 \hbar \omega = \frac{B \langle W \rangle}{A + 2B \langle W \rangle} N \hbar \omega = \frac{\langle W \rangle}{W_S + 2 \langle W \rangle} N \hbar \omega \qquad (4.24)$$

Once the laser action stops, these atoms release this energy through the process of spontaneous emission. We now analyze the dynamics of spontaneous emission.

4.3 Life-Time and Amplification

We have described three different processes involved in the interaction between a two level atom and light. We may get the wrong impression that stimulated processes are continuous in time, while the spontaneous emission is an abrupt process which spoils this continuity. But this is not correct. Spontaneous emission is also a continuous process according to Einstein's picture. What are the dynamics of spontaneous emission? We can deduce them by looking at equation (4.13) when there is no external field, i.e. $\langle W \rangle = 0$. We have

$$\frac{dN_2}{dt} = -A_{21} N_2 \qquad (4.25)$$

Therefore,

$$N_2(t) = N_2(t = 0) e^{-A_{21} t} \qquad (4.26)$$

The life-time is thus $\tau = 1/A_{21}$, i.e. it is inversely proportional to the rate of spontaneous emission. Therefore the population decreases exponentially. This is the same as in any process where the rate of change of some quantity is proportional to that quantity (e.g. nuclear decay).

We now turn to trying to describe the process of amplification of light. For amplification it is necessary that the rate of stimulated emission is much bigger than the rate of spontaneous emission. We will explain this in more detail later, but the basic reason for this is that stimulated emission is coherent and in the same direction as the original light stimulating it and this is the light that actually makes up the laser output. Spontaneous emission is, on the other hand, a random process: photons are scattered randomly in all directions and do not contribute constructively to the laser signal. This requirement leads to

Amplification Criterion

$$\frac{A_{21}}{B_{21} u(\omega_{12})} = e^{\hbar \omega_{12}/kT} - 1 \ll 1 \qquad (4.27)$$

Let's look at two different regimes: the microwave and the visible light at the room temperature ($T = 300K$). For $\lambda = 0.1m$, the value is close to 0, while for $\lambda = 500nm$ it is huge: e^{100}.

So, we have reached a very embarrassing conclusion at the beginning of the book. It looks as if light amplification is impossible in the visible domain — ergo, lasers are impossible! Masers (microwave radiation amplification by stimulated emission), on the other hand, seem to be possible according to this criterion. However, a closer look at amplification reveals that this is also impossible. The ultimate condition for amplification is the achievement of population inversion between two levels, i.e. $N_2 > N_1$. This **Population** will follow rigorously from the study of propagation of light in a **Inversion** medium, which we will do in the next chapter. In thermal equilibrium, population inversion is, of course, impossible as the weights of states go as $e^{-\beta E}$ so that N_2 is always less populated than N_1. However, we have just shown that the steady state rate for N_2 is given by

$$N_2 = \frac{\langle W \rangle}{W_S + 2\langle W \rangle} N \qquad (4.28)$$

which is always less than $N/2$ (it approaches this limit for high intensities as $\langle W \rangle \to \infty$). So the population inversion is impossible not only in thermal equilibrium, but also under the presence of an external coherent source, and this conclusion is independent of the frequency of radiation. Thus, we have used Einstein's theory of the light–matter interaction to conclude something that is obviously wrong, namely that amplification of light is impossible. But lasers do exist. So, we need to be smarter in our reasoning. In order to do so we need to review some basic optical processes first.

Final Thought: We have now been able to describe a great deal of how light behaves, by using the fact that the blackbody radiation is quantized. We can obtain some vary basic facts about the light–matter interaction and can also describe the dynamics of this process in a very heuristic way. As things become more and more complicated, we will see that this very basic (almost) classical treatment becomes less and less adequate. We will then start to approach the fully quantized model. The main result to pay attention to so far is Einstein's derivation and the interaction of a two level atom with radiation within this framework.

Chapter 5

Basic Optical Processes — Still Classical

We saw that Newton thought that light was made up of particles. Then, it was gradually discovered that it is, in fact, a wave. This culminated in Young's double slit experiment and the derivation of the wave equation from Maxwell's four equations of electricity and magnetism. So, light behaves like a wave, and therefore must display effects such as interference and diffraction.[1] This last statement should, of course, be phrased the other way round from the experimental perspective: light can be seen to diffract and interfere, and therefore it must be a wave! Let's look some more into the phenomenon of interference.

5.1 Interference and Coherence

Basic interference processes have already been mentioned at the beginning of the book. Interference is one of the key properties in understanding light classically (as well as quantumly) and its manifestation at the level of individual systems is what makes quantum mechanics counterintuitive and exciting. In general, interference means that the effect from a number of individual sources of disturbances (electromagnetic waves, for example), is larger or smaller than the sum of individual effects (i.e. they either constructively or destructively interfere).

Suppose that we have N atoms, each emitting a wave of light. A typical wave emitted would mathematically be described by

$$\Psi_k = a_k e^{i(\omega t + \delta_k)} \tag{5.1}$$

where a_k is the (time independent) amplitude of the kth wave.

---

[1]With the advent of quantum mechanics, light again become a particle, but a weird one indeed. It is a particle that behaves like a wave! Quantum mechanically, everything is a particle behaving like a wave, so as our description becomes more weird, it also becomes more unifying at the same time.

Let's now add up all the contributions to see the total amplitude and then take the square to obtain the total effect (intensity if you like)[2]

$$I = \sum_i a_i^2 + \sum_k \sum_{j \neq k} a_k a_j \cos(\delta_k - \delta_j) \qquad (5.2)$$

When atomic emission is not co-ordinated (which is usually the case unless there is some care taken in order to co-ordinate it), then the phases vary randomly and in the end the "cos" term averages out to zero. Therefore,

$$I = \sum_i a_i^2 = N\overline{a^2} \qquad (5.3)$$

So this is the same as if there was no interference, as the total effect (intensity) is the same as the sum of individual effects (intensities). For example, if we are talking about an ordinary lamp (or a candle), atoms in the lamp's filament are completely independent from one another when emitting light. They will all emit at different times and the phases will all cancel out in exactly the way described above.

Suppose, however, that we could make sure that all the phases are the same (difficult for a conventional source such as a candle or a lamp, but possible in the case of the laser). Then, from equation (5.2)

$$I = N^2\overline{a^2} \qquad (5.4)$$

and this is N times more intense than when the phases are random. Technically speaking, the sources in this case are said to be *coherent*. Since light is a three-dimensional spatial wave evolving in time, there are two different coherences we have to think about:

- Temporal coherence concerns the same "beam of light" at the same spatial point, but two different times. So, typically we would like to look at the relationship between the electric fields $E(r, t_1)$ and $E(r, t_2)$.
- Spatial coherence, on the other hand, concerns two or more different points of the wave, but at the same time.

Coherence is therefore very important to maintain in order to observe interference, so let's discuss it a bit more. The best way to understand temporal coherence is to consider the Mach–Zehnder interferometer. We assume that when we use a quasi-monochromatic

---

[2]This addition is formally known as *interference*.

light, i.e. the spread in frequency of light involved, $\delta\nu$, is small in comparison with the frequency of light itself (this is true for laser light as we have seen). The beam is first split by a beam splitter and then recombined at the same beam splitter after the two "branches" have travelled different distances. If the time delay between the two beams is δt, then it is a well-documented experimental fact that the fringes will be formed *only if*

$$\delta t \delta \nu < 1 \qquad (5.5)$$

(This is nothing to do with the uncertainty relations and does not contradict them — think about it!) This is a manifestation of temporal coherence where the same beam at two different times interferes with itself. The explanation of the above formula lies in the fact that if the time difference between the two beams is larger, then the frequencies of individual components will interfere destructively and there will therefore be no visible fringes. Each beam will have a number of frequencies present and each one will lead to a spatial oscillation leading to a periodic structure. If the delay between the two beams is too big, then the maxima from the corresponding frequency contributions in the two arms will be further and further apart from each other. This periodic intensity distribution will then be very much out of step, leading to washing out of fringes. For a typical thermal source (lamp), $\delta\nu \approx 10^8 s^{-1}$, so that $\delta t \approx 10^{-8} s$. The spatial coherence length is $\delta l = c \times 10^{-8} s = 3$ meters. For a typical laser, on the other hand, $\delta\nu \approx 10^4 s^{-1}$, so that $\delta l = 30 km$.

Spatial coherence, on the other hand, is understood by analyzing Young's double slit experiment. Here the same light wave illuminates two slits "close" to each other. Let the source of light have the size δs, and the angle between the source and the two slits be $\delta\theta$. Then, the interference fringes will form only if

$$\delta\theta\delta s < \bar{\lambda} \qquad (5.6)$$

The explanation of this equation is the same as that for temporal coherence. The source is composed of many point sources and each one of them will lead to a different fringe pattern. Since all the point sources are out of phase with each other we will have to add all the "point" intensities at the end to obtain the total fringe pattern. If the distance between the slits increases (i.e. $\delta\theta$ increases), then the individual patterns become more and more out of step, leading to the disappearance of fringes. This is the basic idea behind the concept of spatial coherence. You can read in Mandel and Wolf [Mandel and Wolf (1995)] much more about the details of spatial and temporal coherence if you are interested in them.

5.2   Light Pressure

Let's first revise your knowledge of radiation pressure. I will do a back-of-an-envelope calculation that can be made more rigorous and leads to the same conclusion. First of all, why are we doing a classical calculation of the pressure when we know that its origin is due to photons colliding with a material and bouncing off or being absorbed which transfers momentum to that material? Again, there are two reasons. First, the classical calculation is very instructive and every physicist should know it! More importantly, the quantum calculation produces exactly the same result — so there is no need to complicate our life at this stage.

Here is how it goes. As you know well, a charge q interacts with the electric field $\mathbf{E}$ through the force given by

$$\mathbf{F}_E = q\mathbf{E} \tag{5.7}$$

and with the magnetic field $\mathbf{B}$ through the Lorentz force

$$\mathbf{F}_H = q\mathbf{v} \wedge \mathbf{B} \tag{5.8}$$

where $\mathbf{v}$ is the velocity of the charge. I will drop the vector notation since it is clear which quantities are vectors and which are not. It is the combined effect of these two forces that creates radiation pressure. The pressure is the force on an area A per that area. Force in turn is the rate of change of momentum, so

$$P = \frac{\frac{\Delta p}{\Delta t}}{A} \tag{5.9}$$

where P is the pressure and Δp is the change in momentum.[3] The momentum density is given by the Poynting vector, S, divided by c^2. The volume perpendicular to the area A is $c\Delta t$. Thus,

Radiation Pressure

$$P = \frac{(S/c^2)cA}{A} = \frac{S}{c} = \frac{I}{c} \tag{5.10}$$

and so the pressure is intensity divided by the speed of light.[4] From the relationship between intensity and energy density, $I = cu$,

---

[3] Here I will use the capital P for pressure and the small case p for momentum.

[4] Note that this can have a factor of two in front or not depending on whether the surface is reflecting or absorbing respectively.

we can conclude that the pressure of light is equal to the energy density,

$$P = u \tag{5.11}$$

This formula has a strong "thermodynamic" flavor. It looks like an equation of state of a photon gas (something like $pV = 2U/3$ for an ideal gas, where U is the internal energy), and it can certainly be thought of in this way. In fact, if we don't consider just normal incidence, but isotropic radiation in a cavity, then there is a factor of $1/3$ present in the photon gas as well, i.e. $P = 1/3u$.[5] However, from Einstein's relativistic equations we have that the energy density is equal to $u = \rho c$. The pressure is therefore finally given by

$$P = \frac{1}{3}uc \tag{5.13}$$

in the agreement with the main text.

We will now see how the concept of radiation pressure can be understood using Einstein's rate equations, i.e. in a more microscopic manner. Here the concepts of spontaneous and stimulated emission become crucial. The photons that make up an electromagnetic wave of wave vector $\mathbf{k}$ each carry a momentum $\hbar\mathbf{k}$. The Compton effect is a beautiful demonstration of the existence of the photon momentum. (Note that in a medium of refractive index η, the photon momentum is given by $\eta\hbar\mathbf{k}$ — you can think of light moving slower by the amount η which is why it carries less momentum). Suppose now that a photon is absorbed by an atom of mass M. It then gains the velocity of $\hbar k/M$. If the atom then decays we have two possibilities:

- *Stimulated emission.* The emitted photon carries away the momentum $-\hbar\mathbf{k}$ in the same direction as the original photon, so that there is no net momentum transferred to the atom.
- *Spontaneous emission.* Direction of the momentum of the emitted photon can be anywhere within a 4π solid angle (uniform,

---

[5]Here is how the thermodynamic argument goes: the pressure of an ideal gas according to kinetic theory is given by $nm\langle v^2 \rangle$, where n is the number density of the gas, m its molecular mass and $\langle v^2 \rangle$ its mean square molecular velocity. In the case of photons, the product nm is the number density of photons, ρ, and the mean square velocity is equal to c^2, c being the velocity of light. Therefore the pressure is equal to

$$P = \frac{1}{3}\rho c^2 \tag{5.12}$$

random distribution). The atom therefore recoils in some random direction. On average, there is no cancellation of the momentum previously gained as the net spontaneous emission transfer averages to zero.

So in a cycle of absorption and emission we have the net transfer

$$\Delta p = \hbar \mathbf{k} \tag{5.14}$$

from the photons to the atoms, in the direction of the incident beam. This is true even if the process has not reached a steady state. So, the momentum transfer to atoms gives rise to radiation pressure of light.

Consider now a situation where the number of atoms N is large enough to produce small time dependencies in the atomic populations and that $\mathbf{P}$ is the total atom momentum. The rate of change of $\mathbf{P}$ in the presence of radiative energy density $\langle W \rangle$ at a frequency ω (resonant with the ground and the single excited state of the atoms) is proportional to the difference between the absorbed and the stimulated emission rates, like so:

$$\frac{d}{dt}\mathbf{P} = (N_1 - N_2)B\langle W \rangle \hbar \mathbf{k} \tag{5.15}$$

Note that the rate of change of momentum is always positive in equilibrium; a negative rate of change can be obtained if we have a population inversion, $N_2 > N_1$ (which is what we need for lasing in the first place). The number of atoms in the upper level has the steady state value

$$N_2 = \frac{NB\langle W \rangle}{A + 2B\langle W \rangle} = \frac{\langle W \rangle}{W_S + 2\langle W \rangle}N \tag{5.16}$$

$$N_1 = \frac{W_S + \langle W \rangle}{W_S + 2\langle W \rangle}N \tag{5.17}$$

and therefore

$$\frac{d}{dt}\mathbf{P} = \frac{NBW_S\langle W \rangle}{W_S + 2\langle W \rangle}\hbar \mathbf{k} \tag{5.18}$$

For strong fields we have that $W_S \ll \langle W \rangle$ and so

$$\frac{d}{dt}\mathbf{P} = \frac{1}{2}NBW_S\hbar \mathbf{k} \tag{5.19}$$

$$= \frac{1}{2}NA\hbar \mathbf{k} \tag{5.20}$$

Saturation Momentum Transfer Rate

This is the saturation value for transfer rate when we use very strong light beams. Once the saturation regime has been reached, any increase of beam strength produces very little change in the momentum transfer rate.

The steady state momentum transfer rate is equal to the force acting on each atom:

$$\mathbf{F}_{\text{steady state}} = \frac{B\langle W\rangle W_S}{W_S + 2\langle W\rangle}\hbar\mathbf{k} \qquad (5.21)$$

How can we measure this force? I will now describe an experiment that can be used not only to measure the force [Loudon (1964)], but also to separate atoms of different isotopes. Suppose that a beam of atoms is passed perpendicularly to a strong laser light. Then the atoms in the beam interact with light and absorb and emit radiation. The deflection occurs when atoms absorb light followed by spontaneous emission, in which case they gain $\hbar k$ momentum on average perpendicular to their direction of motion. The resulting deflection is typically 10^{-5} radians. Now the two isotopes will have different transition frequencies in general. Suppose that we tune the laser to the transition resonant with one, but not the other isotope. Then only the on-resonant atoms will be deflected, and therefore the two isotopes would separate into two different beams.

Final Thought: The last few sections carry an important lesson. We have seen that the same physical phenomenon can be analyzed from many different angles and at many different levels. Radiation pressure, for example, can be derived using Maxwell's equations and Newton's laws or by applying the basic arguments from kinetic theory and thermodynamics. And this gives us a certain level of understanding. But we can do better if we look at it from the microscopic perspective — using Einstein's rate equations, and obtain not only a deeper understanding, but also a more accurate description.

5.3 Optical Absorption

As light propagates through a medium it gets absorbed and re-emitted by atoms, and therefore its intensity due to absorption will in general decrease. We have seen how this happens when all the atoms start in the ground state and then, through their interaction with laser light, they become excited. We will now consider a more general case. Our analysis will, however, still involve a simple heuristic model for the process of absorbing light when atomic levels in the propagating material are assumed to be sharp.[6] The intensity of radiation is given by the following formula:

$$I = uv \qquad (5.23)$$

where u is the energy density of the radiation field and v is the velocity of the field (I am using v since it can be different from the vacuum value c). Now, as the light propagates through the medium its intensity changes as it gets absorbed and re-emitted. For small distances travelled by light, the change in intensity has to be proportional to the length of the medium travelled and also to the intensity itself (i.e. the more light there is in the first place, the more it will get absorbed). Therefore we arrive at a very simple formula,

$$-dI_\nu(z) = K_\nu I_\nu(z) dz \qquad (5.24)$$

where K_ν is the absorption coefficient which is a function of frequency, but also of the medium itself. Some media will be more absorbing and some less. The solution to this is simple,

Beer's Law

$$I(z) = I(0)e^{-Kz} \qquad (5.25)$$

and is known as Beer's law.[7] Thus the intensity changes exponentially as it propagates. Using the fact that $I = vu$ we can express equation (5.24) as follows:

$$-dI = vudzK \qquad (5.26)$$

On the other hand, we know from Einstein's analysis that

$$-dI = u(\nu)B_{12}(N_1 - N_2)dz \qquad (5.27)$$

Thus, equating the two quantities we obtain

Füchbauer–Ladendurg Formula

$$K = \eta \frac{c^2 \pi^2 A_{21}}{\omega^2}(N_1 - N_2) \qquad (5.28)$$

---

[6]This can be derived rigorously from Maxwell's equation where we take the medium polarization into account leading to a modified wave equation:

$$\left(\frac{\partial^2}{\partial z^2} - \frac{1}{c^2}\frac{\partial^2}{\partial t^2} \right) E(t,z) = \frac{4\pi}{c^2}\frac{\partial^2}{\partial t^2}P(t,z) \qquad (5.22)$$

For those interested in this detailed treatment have a look at Allen and Eberly [Allen and Eberly (1975)]. I should add that this perhaps is not the most fundamental understanding from the classical perspective; the most fundamental treatment is fully microscopic, looking at the propagating medium as a collection of oscillating dipoles. We will do this shortly.

[7]Note the same exponential decrease as in spontaneous emission, but now in space rather than time.

where $v = c/\eta$ (I use η here instead of n for refractive index just to avoid confusion). When $N_1 \rightarrow N_2$ then we see that $dI = 0$ and there is no absorption. This is known as saturation. You don't really have to remember all the terms in this equation, but note the presence of spontaneous emission A, frequency ω, and the ratio of the populations N_1 and N_2.

Saturation

For amplification we need $N_2 \gg N_1$. This is the basis of laser operation and has to hold if we are to have a functional laser. This condition, we have seen, cannot be achieved by exciting a two level atom. This is because when we have saturation, the number of atoms in the ground state is larger than the number in the excited state. However, we can achieve this when we have a three level atom.

5.4 Amplification: Three Level Systems

We mentioned that lasing will never be achieved with a two level system as we cannot obtain the population inversion. How about if we add another level, will that help? The answer is yes, and the reason is very simple. We can store the atom population in this third level from which it will then decay slowly to the second level. From the second level it will lase quickly to the first level from which we will stimulate it back to the third level. And this seems like something that could provide a good model for a realistic laser. Let's write down the rate equations for all the three levels to see all the processes present in the model.

$$\frac{dN_1}{dt} = N_3 B_{13} u(\omega_{13}) + N_3 A_{31} + N_2 B_{12} u(\omega_{21}) + N_2 A_{21}$$
$$- N_1 B_{13} u(\omega_{31}) - N_1 B_{12} u(\omega_{21}) \qquad (5.29)$$

The terms in the above equation should be self-explanatory: for instance, the first term represents stimulated emission from the level 3 to the level 1, i.e. the number of atoms arriving to level 1 from level 3 per unit time and per unit volume. Please go through all the terms and make sure that you understand their meaning. Similar equations hold for the other two levels:

$$\frac{dN_2}{dt} = N_3 S_{32} - N_2 A_{21} + B_{12} u(\omega_{21}) N_1 - B_{12} u(\omega_{21}) N_2 \quad (5.30)$$

$$\frac{dN_3}{dt} = N_1 B_{13} u(\omega_{31}) - N_3 B_{13} u(\omega_{31}) - N_3 A_{31} - N_3 S_{32} \quad (5.31)$$

Note that the constant S represents the transition probability due to both radiative and non-radiative processes (it's just convenient to bunch them together to save some space). Finally, since the total number of atoms is conserved, we have that

$$\frac{d}{dt}(N_1 + N_2 + N_3) = 0 \tag{5.32}$$

Solving these equations in general is tedious, but fortunately we need not do so for our purposes. Recall that we want to use levels 1 and 2 for lasing, so we can solve the above equations in the steady state case (i.e. $dN_i/dt = 0$) to obtain

$$N_2 - N_1 = \frac{\frac{NB_{13}u(\omega_{31})S_{32}}{B_{13}u(\omega_{31}) + A_{31} + S_{32}} - A_{21}}{2B_{12}u(\omega_{21}) + A_{21} + \frac{B_{13}u(\omega_{31})S_{32} + B_{13}u(\omega_{31})(B_{12}u(\omega_{21}) + A_{21})}{B_{31}u(\omega_{31}) + A_{31} + S_{32}}} \tag{5.33}$$

(prove this result — it is not at all easy and should take a few pages of calculations), where

$$N = N_1 + N_2 + N_3$$

is the total number of atoms in all three levels. The number of atoms arriving at the level 2 per unit time is

$$N_3 S_{32} = \frac{NB_{13}u(\omega_{31})S_{32}}{B_{31}u(\omega_{31}) + A_{31} + S_{32}} \tag{5.34}$$

Therefore the probability that an atom arrives at 2 is

$$\Omega = \frac{B_{13}u(\omega_{31})S_{32}}{B_{31}u(\omega_{31}) + A_{31} + S_{32}} \tag{5.35}$$

We now use this to express the numerator of $N_2 - N_1$ (we don't need to bother about the denominator as it is positive at all times). It is

$$N(\Omega - A_{21}) \tag{5.36}$$

The condition for lasing is that we can achieve population inversion, so that $N_2 \geq N_1$. Thus,

$$\Omega > A_{21} \implies \frac{B_{13}u(\omega_{31})S_{32}}{B_{31}u(\omega_{31}) + A_{31} + S_{32}} > A_{21} \tag{5.37}$$

Population Inversion — Three Level Systems

So, lasing can be achieved with a three level atom and the inversion threshold pumping rate is given by the spontaneous emission of the level 2 to the level 1. It is obvious from this formula that a two level atom lasing is impossible as the left hand side of the above is zero while the right hand side is positive.

Final Thought: So in our program of trying to understand the existence of lasers in the lab, we have seen that a two level atom cannot properly explain the population inversion phenomenon which is crucial for the action of the laser. Nevertheless, this is not a serious drawback, since we can always imagine a different process, such as some form of collision pumping, where the atoms are initially prepared in the excited state. On the other hand, a three level atom can fully explain the population inversion without relying on any such external factors.

We have now been talking about emission and absorption in a more or less heuristic way, where the atom had to be quantized (i.e. had a discrete number of levels). Can classical physics explain these processes from the microscopic perspective without involving any additional "quantum" assumptions? Amazingly enough, the answer is yes, and to a pretty high degree of accuracy in certain cases. We turn to this next.

5.5 Classical Treatment of Atom-Light Interaction

One of the best analogues to understand the interaction between radiation and matter is to imagine a pendulum which represents the atom, and a string in the neighborhood of the pendulum representing the field. The two can oscillate separately without affecting each other. Suppose that at the beginning the pendulum is stationary and the string is vibrating; if we couple the string with a piece of thread, then the initial oscillations of the string will be transmitted to the pendulum and the pendulum will start to oscillate as well at the same frequency. If the frequency of the pendulum is very different, then the oscillations will be very small. If the two have similar frequencies, then the oscillations will be large, and all the energy of the string will be transferred to the pendulum — this corresponds to atom absorbing a photon. After some time the energy will go back to the string which is the reverse of the previous process, corresponding as it does to the emission of a photon by the atom. This is the way that atom-radiation interaction is understood classically, but also happens to be true quantum mechanically — you just have to quantize the pendulum and the string! This is basically Dirac's model of the interaction between radiation and matter, where we have individual Hamiltonians for the radiation field and the atom and, in addition, a perturbation Hamiltonian representing their interaction. But we will do this in much more detail later on. Let's still remain classical for some time.

5.5.1 *Dipole radiation*

Now we consider a very simple model of atom–light interactions which is fully classical. The atom is represented as a mass (electron) on a spring (attached to the nucleus if you like). This spring is then contracted and extended as it interacts with light — which is an electromagnetic wave. As the spring extends the energy from the electromagnetic field gets stored in it (corresponding to absorption of radiation) and is then released when the spring contracts (this corresponds to radiation emission). So, let's write Newton's second law for this situation

Driven Harmonic Oscillator

$$\frac{d^2x}{dt^2} = -\omega_0^2 x - \frac{F}{m} \tag{5.38}$$

where F is the force on the electron due to the field $F = qE_0 \cos(\omega t)$. (Note that as usual we neglect the magnetic side of the light wave as its effect is negligible). We can solve this simple differential equation using a simple ansatz $x = x_0 \cos \omega t$. The solution obtained is

$$x = A \cos(\omega_0 t - \phi_0) + \frac{qE_0}{m(\omega^2 - \omega_0^2)} \cos \omega t \tag{5.39}$$

where ϕ_0 is the initial phase of the oscillator and A is a constant. Prove this result by substituting it back into the equation and confirming its validity.

So we see two things. First, the atom oscillates at the frequency of the driving field. Second, the highest amplitude of oscillation is when the field is on resonance, i.e. when the driving frequency is the same as the natural oscillator frequency. The special status of the on resonance driving will be even more pronounced in the quantum treatment later on. Now, the fact that classically the light that is emitted by an oscillating atom is of the same frequency as the one that is absorbed is very awkward when it comes to explaining Compton's experiment. In this experiment light is bounced of a stationary electron so that after the collision the electron starts to move and light changes its direction. Since there is a transfer of momentum from light to the electron, light loses energy and therefore changes its wavelength (which becomes larger). But this change cannot be accounted for classically in any way. Here, therefore, we reach a turning point. We need to use a more sophisticated approach to understand this, which will be done in Chapter 8.

5.5.2 *Radiation damping*

What about the damping? You will remember that there is usually another term in the equation of motion, since the oscillator will be in some kind of medium which will be "viscous" and will provide some resistance to the oscillator's motion (the origin of this apparent viscosity will be explained below). The oscillator (e.g. the atom) will therefore lose energy as it oscillates and will eventually stop oscillating. This, interestingly enough, leads to a line width of the "transition frequency" of the oscillator. This means, among other things, that atomic lines as a consequence will no longer be sharp, but can be assumed to have a certain spread. The correct equation in this case is

Damped Oscillator

$$\frac{d^2x}{dt^2} + \gamma\frac{dx}{dt} + \omega_0^2 x = 0 \tag{5.40}$$

We assume that there is no external driving here. The solution is given by (prove it)

$$x(t) = e^{-\gamma t/2}(Ae^{i\omega_0 t} + Be^{-i\omega_0 t}) \tag{5.41}$$

(This is in the case of the so-called "light damping" so that we have $\omega_0 \gg \gamma$ — see if you can derive this.) This solution decays exponentially to zero. We expect something like this from damping — all oscillations must eventually die away as energy dissipated into the environment. Let's set one of the amplitudes to zero in order to simplify our analysis (say $B = 0$). The resulting solution is not a monochromatic wave, i.e it has more than one frequency component present in its expansion. In order to show this we look at its Fourier spectrum by taking a Fourier transform of it and obtaining

$$\phi(\nu) = \int_0^\infty e^{-\gamma t/2} Ae^{i\omega_0 t} e^{-2\pi i\nu t} dt \tag{5.42}$$

Ask yourself why zero is the lower limit. This integral can simply be evaluated to yield (show it)

$$\phi(\nu) \propto \frac{1}{2\pi i(\nu_0 - \nu) - \gamma/2} \tag{5.43}$$

where $2\pi\nu_0 = \omega_0$. So there are many frequencies now present in the spectrum and not just that of the driving field. The intensities of various frequencies are given by $I(\nu) = |\phi(\nu)|^2$ which leads to the formula for Lorentzian broadening plotted in Figure 5.1. The formula is

$$I(\nu) \propto \frac{1}{4\pi^2(\nu_0 - \nu)^2 + \gamma^2/4} \qquad (5.44)$$

which is plotted in Figure 5.1.

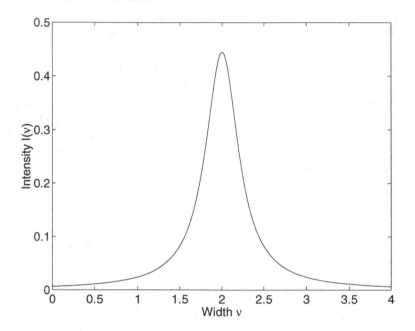

Fig. 5.1 A typical Lorentzian profile of a broadened atomic transition line. ν_0 is equal to 2 and γ is equal to 3 in this graph.

Again the "on resonance" driving is important as it leads to the maximum in intensity — something that will be confirmed quantum mechanically later on. The interesting point is that the width of the transition is proportional to the damping (this is basically the same as the rate of spontaneous emission, i.e. Einstein's coefficient A, but remember that this is a purely classical treatment and we shouldn't really use this concept now).

Where does the damping come from physically? The original argument was due to Lorentz. It requires a very advanced knowledge of classical electrodynamics. Here we follow a beautiful alternative argument proposed by J. J. Thompson. It goes like this. When a charge is stationary the electric field lines originate from the place where the charge is and stretch away to infinity. When the charge is accelerated by Δv in time Δt, then the signal for the change of the field lines will only propagate within a sphere of radius $c\Delta t$ as it propagates at the speed of light. So the lines outside the sphere won't match the ones within as they haven't yet received the signal that the charge has been accelerated in the first place. But this cannot be since the field lines need to be continuous (otherwise the solution to Maxwell's equations would be non-physical). This means that at the boundary the E field cannot just have a radial component, but must also have a tangential component. This, in turn, means that there is a component of the electric field which is propagating away from the charge (Poynting's vector, which defines the direction of propagation, is perpendicular to this component) and taking away some energy. Finally all the energy is taken away in this manner and so the atom stops oscillating. This is the origin of the damping force mentioned before. Using simple trigonometry (if you skip this, you can still understand the rest of the book) the size of the tangential field can be calculated to be

$$E_\theta = \frac{q\ddot{r}\sin\theta}{4\pi\epsilon_0 c^2 r} \tag{5.45}$$

Here r is the position co-ordinate of the oscillator and θ is the usual azimuthal angle in the spherical polar co-ordinates (there is no ϕ dependence as the problem treated is symmetric around the z axis). From this, we can obtain the rate of energy loss. This rate is equal to the Poynting's vector integrated over the surface of a sphere centered on the dipole. Poynting's vector is in turn given by $E \wedge H = E_0^2/Z_0$, where Z_0 is the impedance of vacuum. We therefore obtain

$$\frac{dU}{dt} = -\int_0^\pi \frac{q^2\ddot{r}^2\sin^2\theta}{16\pi^2\epsilon_0 c^3 r^2} 2\pi r^2 \sin\theta d\theta \tag{5.46}$$

$$= -\frac{q^2\ddot{r}^2}{6\pi\epsilon_0 c^3} \tag{5.47}$$

where $2\pi r^2 \sin\theta d\theta$ is the area element and we have used the fact that

$$\int_0^\pi \sin^3 d\theta = 4/3 \tag{5.48}$$

The minus sign is there as the energy is taken away from the dipole. We can now use this to give a precise value of the damping constant γ. Assume that

$$x = x_0 \cos(\omega_0 t) \tag{5.49}$$

so that

$$-\frac{dU}{dt} = \frac{\omega_0^4 e^2 x_0^2}{6\pi\epsilon_0 c^3} \cos^2 \omega_0 t \qquad (5.50)$$

The average value of $\cos^2(\omega_0 t)$ over a cycle is $1/2$ so that the average loss of energy by radiation is

$$-\frac{dU}{dt} = \frac{\omega_0^4 e^2 x_0^2}{12\pi\epsilon_0 c^3} \qquad (5.51)$$

But, since $U = 1/2 m\omega_0^2 x_0^2$ for a simple harmonic oscillator, we have that

Radiation Damping

$$\frac{dU}{dt} = -\gamma U \qquad (5.52)$$

where the decay rate constant γ is given by

$$\gamma = \frac{e^2 \omega_0^2}{6\pi\epsilon_0 c^3 m} \qquad (5.53)$$

Therefore the energy of the oscillator decays away at the above rate, which in turn justifies the usage of the damped oscillator equation to describe atomic interaction with light. This is, as we have indicated, the classical analogue of the quantum concept of spontaneous emission. We have seen from Einstein's treatment that we could understand the concept of spontaneous emission in a heuristic manner, but the most fundamental understanding comes from quantum electrodynamics and is well beyond the current level (although we will have a good stab at it when we quantize light in the second half the book).

5.6   Spectral Lines

Atomic states have in principle well-defined energies. This is true classically as well as quantum mechanically. You will remember that by solving the Schrödinger equation you obtained a set of eigenstates for the hydrogen atom Hamiltonian with the corresponding well-defined eigenvalues which represent values of energy in those levels. However, in practice each of these energy levels, when analyzed spectroscopically, will appear to be broadened. So, we will never really record a single sharp line, but there will instead be a range of energies present. Where do these come from? Three main mechanisms for spectral line broadening are:

- *Atomic collisions.* The process of photon emission takes some time. The atom that is emitting can undergo a collision with another atom during this time. This means that the light wave train that is being generated during emission will be interrupted during this collision and the resulting emission is then not a single frequency pulse, but involves a number of different frequencies superimposed together (a Fourier series). Hence we have line broadening due to collisions. This argument is very "hand waving" but it does explain the main feature of the line broadening due to collisions.

- *Doppler broadening.* This occurs because atoms move as they emit (and they all move at different velocities governed by the Maxwell–Boltzmann distribution formula) and hence their lines are Doppler shifted by different amounts depending on the velocity. This adds a Gaussian envelope to every energy present and we derive this in detail below.

- *Natural line width.* Even if the atoms were not colliding or indeed moving, there would still be a line width present. This sounds intriguing as it appears that there are no other mechanisms possible for this. However, spontaneous emission also leads to broadening. The reason for this is that, as we have seen, spontaneous emission makes the oscillations appear to be multichromatic. The "real" reason will come from the full quantum treatment and you have to wait until Chapter 10 to see it fully.

At this stage we are in a position to expand a bit more on the derivation of the Doppler shift natural line width. The Maxwell–Boltzmann velocity distribution of atoms in a gas is given by

$$N(v) = N_0 e^{-mv^2/2kT} \tag{5.54}$$

where $N(v)$ is the number of atoms moving with the velocity v. The line shape is consequently a Gaussian of the form

Doppler Broadening

$$I(\lambda) = I_0 e^{-(\lambda-\lambda_0)^2/a^2} \tag{5.55}$$

where $a^2 = 2\lambda^2 kT/mc^2$. The Doppler broadening gives in general a different line shape to the line broadening due to radiation damping. This is a Gaussian profile, rather than a Lorentzian profile as in equation (5.44). Finally, I'd just like to show you how to derive the width of the broadening. The Doppler formula is

$$\omega' = \omega \sqrt{\frac{1-v/c}{1+v/c}} \approx \omega\left(1 - \frac{v}{c} + \cdots\right) \tag{5.56}$$

where the approximation is valid at small velocities and atomic velocities are small compared to the speed of light. Therefore,

$$\Delta\omega = \omega - \omega' = \omega\frac{v}{c} \qquad (5.57)$$

On the other hand, the velocity is given by

$$v = \sqrt{\frac{3kT}{m}} \qquad (5.58)$$

thus leading to the formula for frequency broadening.

Final Thought: We have now explained as much of atom–light interaction as we can using the classical approach. This is a pretty sophisticated theory, which can explain a lot of things, but is unfortunately wrong. We have to eventually move onto the next higher level, where we use a more accurate, quantum mechanical treatment to analyze light. The two most important results in this part are the achievement of population inversion and hence lasing by a three level atom, and the classical treatment of radiation and line broadening.

Chapter 6

More Detailed Principles of Laser

Lasers consist of, as we indicated a number of times so far, just a cavity which contains a lasing medium (for simplicity treated as a bunch of two or three level systems but, of course, we can have higher dimensionality). We now want to take all the parameters into account, such as the cavity mirror reflection and the broadening of the atoms inside, to derive a more detailed condition for lasing (and not just the population inversion condition which was obtained from a very simplistic analysis). Every laser has four main elements which are necessary for its successful operation:

- a cavity with two or more highly reflecting mirrors (many different shapes of mirrors are studied in general, and for different purposes; this is, however, well beyond the scope of the present treatment);
- a gain medium which can support an inverted atomic population between the lasing levels;
- an energy source which can excite the atoms in the gain medium to achieve the population inversion;
- a loss mechanism by which the stored energy is dissipated.

We will now review some basic theory behind laser operation.

6.1 Basic Theory: Classical Electrodynamics

Let's again look at absorption and amplification in a certain medium, but now not from the bulk point of view. When an electromagnetic field propagates through a medium, the medium reacts to it. Focusing only on the electric field, the response of the medium is summarized in a quantity called polarization. Imagine that the medium is composed of a bunch of classical oscillating springs (as before); the microscopic dipole is then

$$\mathbf{p} = -e\mathbf{x} \qquad (6.1)$$

and the polarization, $\mathbf{P}$, is defined as

$$\mathbf{P} = N\mathbf{p} = -Ne\mathbf{x} \qquad (6.2)$$

Therefore the driven harmonic oscillator equation for polarization is

$$\ddot{\mathbf{P}} + \gamma\dot{\mathbf{P}} + \omega_0^2\mathbf{P} = \frac{Ne^2}{m}\mathbf{E} \qquad (6.3)$$

We solve this equation as usual by assuming solutions of the form

$$E = E_0 e^{i(\omega t - kz)} \qquad (6.4)$$
$$P = P_0 e^{i(\omega t - kz)} \qquad (6.5)$$

meaning that P depends on E linearly. We therefore obtain

$$\mathbf{P} = \frac{Ne^2}{m}\frac{1}{\omega_0^2 - \omega^2 - i\omega\gamma}\mathbf{E} \qquad (6.6)$$

Note that the proportionality constant between polarization and the electric field is a complex number. This may seem a paradox, until we remember that the quantity we measure is the intensity which is the mod square of the field. The real part of this constant represents the decay of the intensity due to the absorption in the medium, while the imaginary part represents the oscillatory behavior of the field.

To see this, let's continue to assume that the response of the medium is linear so that $P = \epsilon_0 \chi E$ (I'll now drop the bold face vector notation since it's clear which quantities are vectors and which are not), where χ is, as before, the dielectric susceptibility. (No higher powers of the electric field are present — we will look at the generalization to non-linear situations shortly.) From Maxwell's equations we obtain the following wave equation:

$$\nabla^2 E - \frac{1}{c^2}\frac{\partial^2 E}{\partial t^2} = \frac{1}{\epsilon_0 c^2}\frac{\partial^2 P}{\partial t^2} \qquad (6.7)$$

Take E to be a plane wave as before. Then by substituting into the above equation we obtain

$$\left(-k^2 + \frac{\omega^2}{c^2}\right) = -\chi\frac{\omega^2}{c^2} \qquad (6.8)$$

so that the wave vector is

$$k^2 = (1 + \chi)\frac{\omega^2}{c^2} \qquad (6.9)$$

On the other hand, $k^2 = n^2\omega^2/c^2$ where n is the refractive index. Therefore we can conclude that

$$n^2(\omega) = 1 + \chi(\omega) \qquad (6.10)$$

Since χ is a complex number, so is the refractive index! We can write it in terms of its real and imaginary components (the choice of notation will become clear soon):

$$n = n' - i\kappa \qquad (6.11)$$

The real part (n') is the "normal" refractive index, while the imaginary component $(-\kappa)$ is the absorption or gain. To see this remember that whilst in vacuum $c = \lambda\nu$; in a medium we have that

$$\frac{c}{n} = \frac{\lambda}{n}\nu \qquad (6.12)$$

since, as we know, only the wavelength (and not the frequency) changes. Therefore

$$k = \frac{2\pi n}{\lambda} = \frac{2\pi}{\lambda}(n' - i\kappa) \qquad (6.13)$$

The electric field is therefore modified as it propagates. It has the form

$$E = E_0 e^{i(\omega t - (2\pi/\lambda)n'z)}e^{-i(2\pi/\lambda)(-i\kappa)z} \qquad (6.14)$$

The last term is in fact equal to $e^{-(2\pi\kappa/\lambda)z}$, which is an exponential decay. Note that here I am looking at the z component only. The full three-dimensional case would be exactly the same and there would be similar terms for x and y. In terms of intensity this can be written as

$$I(z) = I(0)e^{\alpha z} \qquad (6.15)$$

which proves Beer's law we introduced before (please don't confuse this α with the polarizability). The absorption (gain) coefficient is

$$\alpha = 2k_0\kappa = 2\frac{2\pi}{\lambda}\kappa \qquad (6.16)$$

We can calculate κ and n' in terms of microscopic (atomic) features to be

$$\kappa = \frac{Ne^2}{2\epsilon_0 m} \frac{\gamma\omega}{(\omega_0^2 - \omega^2)^2 + \gamma^2\omega^2} \tag{6.17}$$

$$n' = 1 + \frac{Ne^2}{2\epsilon_0 m} \frac{\omega_0^2 - \omega^2}{(\omega_0^2 - \omega^2)^2 + \gamma^2\omega^2} \tag{6.18}$$

So I'd like now to apply this theory to lasers.

We assume that the laser cavity has two mirrors (this is a one-dimensional analysis, but it contains all the important features of the more general treatment) which have the reflection coefficients r_1 and r_2. This means that the intensity I impinging on a mirror of reflectivity will bounce back with the intensity rI. In addition, the gain due to the medium inside is (as we saw before)

$$I(z) = I(0)e^{\alpha z} \tag{6.19}$$

So, let's imagine that we have a combination of one round trip reflections and the gain coefficient. Writing $r_1 r_2 = e^{-2\beta}$, we obtain the intensity of light after one round trip to be

$$I e^{2\alpha L - 2\beta} \tag{6.20}$$

For oscillations to build up we need $\alpha L > \beta$. Incidentally, if we were really pedantic we could include other losses in β and not just due to leaking mirrors. This, however, is usually unnecessary when we want to give general estimates for lasing. We have seen from before (section on optical absorption equation (5.28)) that

$$\alpha = \eta \frac{c^2 \pi^2 A_{21}}{\omega^2} \left(1 - \frac{N_2}{N_1}\right) \tag{6.21}$$

This formula does not take into account that there is a line broadening, which means that it should be modified by the Lorentzian line shape. However, we will ignore this detail in our treatment as the final estimate for the condition for lasing is still very accurate. The maximum value of the gain coefficient is at $\omega = \omega_0$. We can now derive the threshold population inversion from the condition that $\alpha > \beta/L$, which leads to

Threshold Population Inversion

$$N_2 - N_1 = \frac{\beta\omega_0^2}{\eta L c^2 \pi^2 A_{21}} \tag{6.22}$$

This implies that β and L must be properly adjusted to obtain the critical inversion.

To gain some more physical insight into this expression and to estimate the necessary size of the population difference $N_2 - N_1$, we can write the amplification condition in a different way. The amplification will be achieved when the rate of emission into a certain cavity mode exceeds the rate of loss of photons from the cavity in that mode (this means that we want to be amplifying radiation within the walls, and not lose it into the environment). The loss happens, of course, as the walls of the cavity are not perfectly reflecting. The number of photons emitted into a mode is

$$\frac{N_2 - N_1}{\tau \frac{\omega^2 \Delta \omega}{\pi^2 c^3}} \tag{6.23}$$

where $\tau = 1/A_{12}$. The rate of loss of photons is $1/t_1$, where $t_1 = L/(1-r)\eta c$ is the life-time of a photon within the laser cavity and r is the reflectivity of the mirrors (assume the same for both mirrors). The amplification condition now leads to

$$\frac{N_2 - N_1}{\tau} > \frac{V \frac{\omega^2 \Delta \omega}{\pi^2 c^3}}{t_1} \tag{6.24}$$

Let's plug in some numbers to see how big this should be. We have

$$\omega = 10^{15} Hz \tag{6.25}$$
$$c = 3 \times 10^8 ms^{-1} \tag{6.26}$$
$$\tau = 10^{-8} s \tag{6.27}$$
$$\Delta \omega = 10^{10} s^{-1} \tag{6.28}$$

These lead to (assuming that $\eta \approx 1$ and $r \approx 0$)

$$\frac{N_2 - N_1}{V} > 10^{10} cm^{-3} \tag{6.29}$$

which is very different from the "zero" that comes out from Einstein's simplistic analysis (but still not too big compared to the number of atoms which is of the order of 10^{23} per meter cubed).

Final Thought: We have now derived a more rigorous condition for laser operation which takes into account the width of atomic transitions as well as the possibility of atoms leaking out of the cavity. This gives us a condition for lasing that is more stringent than the condition we have from the Einstein balance equation of a two level system interacting with radiation. Nevertheless, Einstein's treatment still offers a fast rule for obtaining a necessary condition for a working laser, namely that we must be able to obtain a population inversion by some means. Still, radiation in a cavity will be a mixture of many different frequencies and they will all oscillate with different phases. The question is why don't all these interfere destructively, canceling each other out and leaving no signal? Well, the answer is that it seems that we will need to look deeper into these issues to gain a better understanding. This is what we will do next, when the wave nature of light will become important again.

6.2   Mode-Locking

In its natural state the laser cavity contains radiation which consists of infinitely many different modes all oscillating at different frequencies (rates). They also have different phases, as the atoms in the cavity mirrors and the lasing medium radiate randomly at different times and therefore contributing different initial phases to the total oscillating field. Thus the state of the field can be written as

$$E(t) = A \sum_{n}^{N} e^{i(\omega_n t + \phi_n)} \tag{6.30}$$

We can now compute the intensity of radiation to obtain

$$I = NA^2 \tag{6.31}$$

and so intensity is proportional to the number of modes (this is, of course, infinite, but let's ignore this fact here by introducing an upper cut-off, N). The process of mode-locking now means arranging things so that all of the fields in different modes have the same phase

$$\phi_n = \phi \tag{6.32}$$

Mode-Locking Condition

We know from before that the frequency separation between different modes is given by

$$\Delta\omega = \frac{\pi c}{L\eta} \tag{6.33}$$

Therefore the mode-locked intensity is now

$$I_{\text{mode-locked}} = A^2 \frac{\sin^2 \frac{n\pi ct}{2L\eta}}{\sin^2 \frac{\pi ct}{2L\eta}} \tag{6.34}$$

where the peak intensity is given by

$$I = A^2 N^2 \tag{6.35}$$

and it is N times stronger than without the mode-locking — hence this procedure. The maximum happens when

$$t = \frac{2mL\eta}{c} \qquad (6.36)$$

The time between the maxima can also be found to be

$$\Delta t = \frac{2L\eta}{c} \qquad (6.37)$$

which is the time of one round trip in the cavity, as one would expect. The purpose of mode-locking can be summarized in one sentence: if modes can be locked, then high intensities and small pulse widths can be achieved if we could get sufficiently many modes to oscillate in phase. And so what about the width of the pulse? It is given by

$$\delta t = \frac{2\pi}{N\Delta\omega} \qquad (6.38)$$

and this relationship allows us to achieve shorter and shorter pulses (known as ultrashort pulses) by "locking in" more and more modes (N).

How can mode-locking be achieved practically? There are many **Practical** ways of achieving this and the methods fall into two main cate- **Mode-** gories: the active and the passive mode-locking. To understand **Locking** the way these methods can be made to work in general let us look at just two modes, ω_0 and ω_1, with the corresponding phases ϕ_0 and ϕ_1. Two (coupled) oscillator systems are ubiquitous in Nature and are at the heart of many phenomena. The oscillators generally oscillate at different rates, and will not naturally oscillate in accord, be in phase. This is the same as the problem of Huygens' pendulums. Huygens was ill in his bed and noticed that two pendulums that he had on his wall opposite to his bed were swinging the same way, i.e. they were completely synchronized. This happened, Huygens realized, no matter how they started initially, i.e. even if they started completely out of phase so that when one was to the left and the other was to the right. Interestingly enough, when he separated them to be on two different walls this no longer happened: if they started out of phase, then they remained out of phase. The explanation that Huygens came up with was that when they were on the same wall then there was a mechanism for them to exchange energy in a coherent way. This means that they were coupled. Interestingly, even this small amount was enough to synchronize the two clocks. When they were separated then there was no mechanism present for them to "communicate" with each other and correlate their phases. So, the moral of Huygens' observation is that we need a coherent coupling between the two modes in a laser. This can be accomplished by taking the mode ω_0 and modulating it so that the form of the field is

$$E_0(t) = E_0(1 + M\cos\omega't)\cos\omega_0 t \qquad (6.39)$$

where M is the so-called modulation depth. This field gives rise to two sidebands which can be seen by writing it in the following form

$$E_0(t) = E_0\cos\omega_0 t + \frac{1}{2}ME_0\cos(\omega_0 - \omega')t + \frac{1}{2}ME_0\cos(\omega_0 + \omega')t \qquad (6.40)$$

We can now choose to make the modulation frequency so that

$$\omega' = \frac{2\pi}{t} = \frac{\pi c}{nL} \qquad (6.41)$$

This is the same as the mode spacing. Then the sidebands coincide with the neighboring modes and this provides a coupling between them in order to correlate their phases. A similar technique can be used on the mode ω_1 so that it couples to the mode ω_2 and so on. There are many different ways of achieving this described frequency modulation. One is to change the losses through the mirrors (active mode-locking) so that they follow the rate

$$k = k_0 + k_1\cos\omega't \qquad (6.42)$$

in which case the new field would be kE. Another way is that this modulation is produced by an absorber inside the cavity, in addition to the gain medium (passive mode-locking).

In conclusion, mode-locking enhances the intensity of light produced by a laser. It is very important to perform light amplification if we want to see light of certain frequency that is not very intense. We turn to this subject next — the non-linear optics and higher harmonic generation.

6.3   Non-linear Optics

What is non-linear optics? The processes we have studied so far involve absorption and emission of radiation by atoms. In Einstein's model, only on resonance absorption is possible, i.e. light has to be of the same frequency as the energy gap between the two atomic levels. However, absorption (emission) is possible even when the photons do not match the atomic transition frequency. For example, an atom of transition frequency ω can absorb two photons each of frequency $\omega/2$. The probability for this process is small, but it is still possible according to quantum mechanics. We will calculate this in more detail when we speak about the perturbation theory and semi-classical treatment of atom–light interactions. However, two photon absorption is, in fact, also possible according to classical physics, i.e. electrodynamics, as we will see in this chapter. There are, of course, no photons in the classical formalism (and atoms are treated as bulk), but we will talk of analogous effects. This is the subject of non-linear optics.

In the classical approximation, non-linear optics reduces to solving non-linear Newtonian equations for an anharmonic oscillator. Consider applying a laser field $E(\omega) \cos \omega t$ to a bound electron. The resulting force is equal to $eE(\omega) \cos \omega t$, where e is the electric charge. This is in most cases much smaller than the Coulomb field that binds the electron to the nucleus, which is 10^{11}V/m. A typical laser generates a peak electric field strength of $10^7 - 10^{10} \text{V/m}$. If this is the case then the response of the medium to the laser field is (very nearly) linear in the sense that the (total) polarization is given by

$$\mathbf{P} = \epsilon_0 \chi^{(1)} \mathbf{E} \tag{6.43}$$

where $\chi^{(1)}$ is the linear susceptibility we had before. What happens if we drive the sample of atoms (electrons) harder, i.e. if we increase intensity of the laser? Then the potential becomes anharmonic. The Taylor expansion of the potential of the electron driving field is

$$V(x) = V(0) + x \frac{dV}{dx} + \frac{1}{2} \frac{d^2 V}{dx^2} + \frac{1}{3!} \frac{d^3 V}{dx^3} + \cdots \tag{6.44}$$

where the second term in the expansion is zero (by definition of equilibrium $dV/dx = 0$), the third term is the harmonic part of the potential and the last term is the first anharmonic part of the potential. How does the medium react to this potential? The polarization is no longer linear and can in general be written as

$$P = \epsilon_0 (\chi^{(1)} E + \chi^{(2)} E^2 + \chi^{(3)} E^3 + \cdots) \tag{6.45}$$

where $\chi^{(n)}$ is the nth (higher) order susceptibility. This leads to the phenomenon of higher harmonics. Consider, for example, the third order term

$$P^{(3)} = \epsilon_0 \chi^{(3)} E_0^3 \cos^3 \omega t \tag{6.46}$$

$$= \epsilon_0 \chi^{(3)} E_0^3 \left\{ \frac{1}{4} \cos 3\omega t + \frac{3}{4} \cos \omega t \right\} \tag{6.47}$$

Therefore we see the presence of a strong driving field — the atomic response leads to the secondary generated field of frequency 3ω, the so-called *third harmonic generation*. If we look at the electron as an oscillator then the resulting equation of motion under the anharmonic perturbation is in general of the form

Anharmonic Oscillator

$$\frac{d^2x}{dt^2} + \omega_0^2 x + \frac{3A}{m}x^2 + \frac{4Bx^3}{m} = \frac{e}{m}E(t) \tag{6.48}$$

If $A = B = 0$ we have the usual harmonic response that we analyzed earlier in the book.

Let's look at the second harmonic generation. In that case we have that $P^{(2)} = \epsilon_0\chi^{(2)}E^2$ where

$$E^2 = E^2(\omega)\cos^2(\omega t - kz) = \frac{1}{2}E^2(\omega)[1 + \cos 2(\omega t - kz)] \tag{6.49}$$

In order to understand this process better, let's think briefly in terms of the full quantum description of the phenomenon. The second harmonic generation actually involves the absorption of two photons of energy $\hbar\omega$ by a two level atom which has an energy difference of $2\hbar\omega$ between the two levels. This atom consequently becomes excited and then emits a photon of the energy $2\hbar\omega$. Therefore two photons at ω have been converted into one photon at 2ω.

Let's try to understand non-linear optics from the microscopic (albeit classical) perspective. Atoms are driven oscillators (we ignore damping for this calculation) and so we need to solve Newton's second law for the anharmonic oscillator. We do this to show the case of the second harmonic generation, although the same kind of calculation (more laborious though) can lead to higher order processes. The dynamical equation is in this case given by

$$\ddot{x} + \omega_0^2 x = -\alpha x^2 \tag{6.50}$$

Now the solution of this equation will involve other frequencies in addition to ω_0 (you can check this by showing that $\cos\omega_0 t$ is no longer a solution of the above equation). In fact the general solution will be some kind of Fourier series, $\sum_n a_n \cos(n\omega_0 t)$, where the coefficients a_n are to be determined. This can be done in a "perturbative" fashion.[1] Assume that

$$x = x^{(1)} + x^{(2)} \tag{6.51}$$

---

[1]Those interested in more detail, and there is much more to it than presented here, are encouraged to consult the excellent text by Landau and Lifshitz [Landau and Lifshitz (1960)].

and that

$$x^{(1)} = a\cos\omega t \qquad (6.52)$$

The frequency can also be expressed as $\omega = \omega_0 + \omega^{(1)} + \ldots$ Substituting this into equation (6.50) and omitting terms above second order we obtain

$$\ddot{x}^{(2)} + \omega_0^2 x^{(2)} = -\frac{1}{2}a^2\alpha - \frac{1}{2}a^2\alpha\cos 2\omega t + 2\omega_0\omega^{(1)}a\cos\omega t \quad (6.53)$$

This equation solved to obtain (the full perturbative treatment can be found in Guenther [Guenther (1990)]):

$$x^{(2)} = -\frac{\alpha a^2}{2\omega_0^2} + \frac{\alpha a^2}{6\omega_0^2}\cos 2\omega t \qquad (6.54)$$

Therefore we see exactly the presence of the frequency twice the size of the fundamental frequency.

Second harmonic generation was experimentally achieved by Franken, Hill, Peters and Weinreich, in 1961 (Phys. Rev. Lett. **7**, 118). If you go and check out the reference you won't see any signal for the second harmonic on the experimental plot in the paper. The reason is that the spot was so small that the editor deleted it thinking it was an error in the preparation of the manuscript for publication!

6.4 Phase Matching

Let's now talk about the dynamics of second harmonic generation. How does a wave propagate in a non-linear medium? The answer lies as always in Maxwell's equations. First we need to change the constitutive relations to allow for non-linear polarization:

$$\mathbf{B} = \mu\mathbf{H} \qquad (6.55)$$

$$\mathbf{J} = \sigma\mathbf{E} \qquad (6.56)$$

$$\mathbf{D} = \epsilon_0\mathbf{E} + \epsilon_0\chi\mathbf{E} + \mathbf{P}_{NL} \qquad (6.57)$$

6.4.1 *Rigorous derivation*

All the quantities in the above equations have their usual meaning. We will assume that the medium is homogeneous and non-magnetic (i.e. the magnetization $M = 0$), which allows us to solve Maxwell's equations to obtain a wave equation in much the same way we did in the first section. Here I mainly follow Guenther's treatment [Guenther (1990)]. The equations we need first are the "curl equations":

$$\nabla \wedge \mathbf{E} = -\mu \frac{\partial \mathbf{H}}{\partial t} \qquad (6.58)$$

$$\nabla \wedge \mathbf{H} = \mathbf{J} + \frac{\partial \mathbf{D}}{\partial t} \qquad (6.59)$$

where $\mathbf{J}$ is the current density and $\mathbf{D}$ is the electric displacement. (The other two Maxwell's equations are $\nabla D = \rho$ and, as always, $\nabla B = 0$). Differentiating the two by dt we obtain

$$\nabla \wedge \frac{\partial \mathbf{E}}{\partial t} = -\mu \frac{\partial^2 \mathbf{H}}{\partial t^2} \qquad (6.60)$$

$$\nabla \wedge \frac{\partial \mathbf{H}}{\partial t} = \epsilon_0 (1 + \chi) \frac{\partial^2 \mathbf{E}}{\partial t^2} + \sigma \frac{\partial \mathbf{E}}{\partial t} + \frac{\partial^2 \mathbf{P}_{NL}}{\partial t^2} \qquad (6.61)$$

and the curl of the same two equations is

$$\nabla \wedge \nabla \wedge \mathbf{E} = -\mu \nabla \wedge \frac{\partial \mathbf{H}}{\partial t} \qquad (6.62)$$

$$\nabla \wedge \nabla \wedge \mathbf{H} = \epsilon_0 (1 + \chi) \nabla \wedge \frac{\partial \mathbf{E}}{\partial t} + \sigma \nabla \wedge \mathbf{E} \qquad (6.63)$$

Combining these two we obtain

$$\nabla \wedge \nabla \wedge \mathbf{E} + \mu \epsilon_0 (1 + \chi) \frac{\partial^2 \mathbf{E}}{\partial t^2} + \mu \sigma \frac{\partial \mathbf{E}}{\partial t} + \mu \frac{\partial^2 \mathbf{P}_{NL}}{\partial t^2} = 0 \qquad (6.64)$$

$$\nabla \wedge \nabla \wedge \mathbf{H} + \mu \epsilon_0 (1 + \chi) \frac{\partial^2 \mathbf{H}}{\partial t^2} + \mu \sigma \frac{\partial \mathbf{H}}{\partial t} = 0 \qquad (6.65)$$

We thus obtain a non-linear wave equation for E, for example, **Non-linear Wave Equation**

$$\nabla^2 \mathbf{E} = \mu \sigma \frac{\partial \mathbf{E}}{\partial t} + \mu \epsilon \frac{\partial^2 \mathbf{E}}{\partial t^2} + \mu \frac{\partial^2 \mathbf{P}_{NL}}{\partial t^2} \qquad (6.66)$$

The non-linearity of this equation is not immediately obvious from the above form. You need to substitute P_{NL} as a function of E in the last term and the resulting equation for E is then non-linear (meaning that a sum of two different solutions of this equation is not another solution of the same equation). So we have arrived at the non-linear wave equation, but how can we understand its physical meaning? Let's look at a one-dimensional mixture of three different frequencies and solve the wave equation for this case

$$E(\omega_i) = E_i = \frac{1}{2}[E_i(z)e^{i(\omega_i t - kz)} + E_i^*(z)e^{-i(\omega_i t - kz)}] \qquad (6.67)$$

We make another natural assumption, i.e.

$$k_i \frac{\partial E_i}{\partial z} \gg \frac{\partial^2 E_i}{\partial z^2} \qquad (6.68)$$

which means that the field varies sufficiently slowly with distance that higher than first order changes can be neglected (in other words, I assume that absorption is not too strong). The left hand side of the wave equation now becomes,[2]

$$\nabla^2 \mathbf{E} = -\frac{1}{2}\left(k_i^2 E_i + 2ik_i\frac{\partial E_i}{\partial z}\right)e^{i(\omega_i t - kz)} + \text{complex conjugate}$$

(6.70)

The crucial term in the wave equation is the last term involving the non-linear polarization. The total polarization due to three different frequencies is a sum of individual polarizations due to single frequencies

$$P_{NL} = P_{NL}(\omega_1) + P_{NL}(\omega_2) + P_{NL}(\omega_3)$$ (6.71)
$$= d[E^*(\omega_2)E(\omega_3) + E^*(\omega_1)E(\omega_3) + E^*(\omega_1)E(\omega_2)]$$

where d is the so-called non-linear optical coefficient (measured experimentally). So the second time derivative of E_1, for example, would be

$$\frac{\partial^2 P_{NL}(\omega_1)}{\partial t^2} = -(\omega_3 - \omega_2)^2 dE^*(\omega_2)E(\omega_3)e^{i(\omega_3-\omega_2)t-(k_3-k_2)z)} + \text{c.c.}$$

(6.72)

The resulting wave equation is now very difficult to solve as the frequencies appear mixed and we cannot separate them (this is because of non-linearity; otherwise, there would be no problem). To simplify the equations we assume

Energy Conservation

$$\omega_3 = \omega_1 + \omega_2$$ (6.73)

After some simple manipulations we can obtain three separate equations, one for each frequency (the full calculation can be found in [Guenther (1990)]):

$$\frac{\partial E_1}{\partial z} = -\frac{\sigma_1}{2}\sqrt{\frac{\mu}{\epsilon_1}}E_1 - \frac{i\omega_1}{2}\sqrt{\frac{\mu}{\epsilon_1}}dE_3E_2^*e^{-i(k_3-k_2-k_1)z}$$ (6.74)

$$\frac{\partial E_2}{\partial z} = -\frac{\sigma_2}{2}\sqrt{\frac{\mu}{\epsilon_2}}E_2 - \frac{i\omega_2}{2}\sqrt{\frac{\mu}{\epsilon_2}}dE_3E_1^*e^{-i(k_3-k_2-k_1)z}$$ (6.75)

$$\frac{\partial E_3}{\partial z} = -\frac{\sigma_3}{2}\sqrt{\frac{\mu}{\epsilon_3}}E_3 - \frac{i\omega_3}{2}\sqrt{\frac{\mu}{\epsilon_3}}dE_1E_2^*e^{-i(-k_3+k_2+k_1)z}$$ (6.76)

---

[2]It is worth remembering here that if $f = Ae^{i(\omega t - kx + \phi)}$ then

$$\frac{\partial^n f}{\partial t^n} = (i\omega)^n f$$ (6.69)

and a similar rule for the spatial derivative.

The term in the exponent is very suggestive as it reminds us of momentum change in this process

$$\Delta k = k_3 - k_1 - k_2 \qquad (6.77)$$

Let's now try to integrate the equation for E_3, for example. To be able to do so we assume that the medium is lossless, i.e. $\sigma = 0$, and that E_1 and E_2 do not depend on z. Then,

$$E_3 = -\frac{i\omega_3}{2}\sqrt{\frac{\mu}{\epsilon_3}}d\int_0^L E_1 E_2 e^{i\Delta kz}dz \qquad (6.78)$$

$$= -\frac{\omega_3}{2}\sqrt{\frac{\mu}{\epsilon_3}}\frac{dE_1 E_2}{\Delta k}(e^{i\Delta kz} - 1) \qquad (6.79)$$

This now allows us to calculate the corresponding Poynting vector, which tells us about the energy (intensity) of the wave. We can see that

$$S_3 \sim S_1 S_2 \frac{\sin^2(\Delta k L/2)}{(\Delta k L/2)^2} \qquad (6.80)$$

So the maximum intensity at ω_3 is obtained either when

Momentum Conservation

- $\Delta k = 0$, which is called the phase matching condition, or
- if $\Delta k \neq 0$, then for $\Delta k L/2 = \pi/4$, i.e.

$$L = \frac{\pi}{2\Delta k} \qquad (6.81)$$

Let's apply this to second harmonic generation. We have that $\omega_1 = \omega_2$ and $\omega_3 = 2\omega_2$. Also, $k_1 = k_2 = n_1\omega_1/c$ and $k_3 = n_3\omega_3 = 2n_3\omega_1/c$. The momentum difference is non-zero and is

Second Harmonic Generation

$$\Delta k = \frac{2\omega_1}{c}(n_3 - n_1) \qquad (6.82)$$

That means that the coherence length is

$$L = \frac{\lambda_1}{4(n_3 - n_1)} \qquad (6.83)$$

In order for a crystal to be an effective generator of second harmonics it should ideally be of this length.

6.4.2 *Heuristic derivation*

To summarize, non-linear responses of a physical medium are seen when the intensity of radiation is high. They are described through the polarizability of the medium, depending not only on the field linearly but also involving higher order terms. In terms of our microscopic picture, atoms in the medium respond as harmonic oscillators, but the potential that they see is now anharmonic, involving terms of cubic and higher order. This can in general be solved, as we did, to show that the atom now responds by oscillating not only at the frequency of the driving field, but also at twice as high frequency, three times as high frequency and so on. This can be understood very simply by observing that if $E = \cos \omega t$, then $E^2 = \cos^2 \omega t = (1 + \cos 2\omega t)/2$, and we immediately see the origin of higher frequencies. The second order, in particular, leads to frequency doubling. Studying these kind of responses is the subject of non-linear optics. The second harmonic response of a medium to the electric field is mathematically described as

$$P_{2\omega} = \epsilon_0 \chi^{(2)} E_\omega^2 \tag{6.84}$$

This is now a source in Maxwell's equations, but let's avoid all the calculations and jump straight to the propagation equation. The field is

$$E_\omega = a_\omega e^{ik_1 z} \tag{6.85}$$

and we assume no depletion as the field is strong, i.e. a is not position dependent. The field at 2ω is, on the other hand, assumed to be affected, so

$$E_{2\omega} = a_{2\omega}(z) e^{ik_2 z} \tag{6.86}$$

The field propagation is now

$$dE_{2\omega} = c P_{2\omega} dz = c\epsilon_0 \chi^{(2)} E_\omega^2 \tag{6.87}$$

where I have bunched all the ϵ's and μ's into the constant c — we won't worry about that now. Solving the "equation of motion" we obtain

$$da_{2\omega}(z) e^{ik_2 z} = c(\epsilon_0 \chi^{(2)} a_\omega^2) e^{2ik_1 z} \tag{6.88}$$

We can integrate this to obtain

$$a_{2\omega}(L) \propto \int_0^L e^{i(2k_1 - k_2) z} dz \tag{6.89}$$

Now we label the phase difference as $\Delta k = 2k_1 - k_2 (= 2\omega(n_1 - n_2)/c)$. The whole point of phase matching is to ensure that the pump and second harmonic don't cancel each other out as we would then lose all the signal and the ability to measure (and use?) the second harmonic. If Δk is large this is exactly what happens. Let's see why. We know that the above integral is just a "sinc" function (i.e. of the form $\sin(x)/x$):

$$\int_0^L e^{i(2k_1 - k_2)z} dz = L e^{i\Delta k L/2} \frac{\sin \Delta k L/2}{\Delta k L/2} \qquad (6.90)$$

The intensity is given by the mod squared of this expression and we have

$$I \propto \sin c^2 \left\{ \frac{\Delta k L}{2} \right\} \qquad (6.91)$$

The maximum is therefore when $\Delta k = 0$, which is known as the phase matching condition. Let's check a few real life numbers:

$$n(\lambda = 694nm) = 1.54 \qquad (6.92)$$

$$n(\lambda = 347nm) = 1.57 \qquad (6.93)$$

This leads to $\sin c^2(2700) = 10^{-7}$, which is a very small number. This is basically why phase matching is important.

I'd now like to discuss one last non-linear effect, the Kerr effect, and this is occurrence of the intensity dependent refractive index. So if the field is sufficiently high then the refractive index starts to depend on intensity and this may lead to self-focusing of the propagating beam, which may destroy the material through which it is propagating. This is a third order process in non-linear optics. To see its origin, remember that the third order non-linear polarization will be given by

$$P_{NL} = \epsilon_0 \chi^{(3)} E^3 = \epsilon_0 (\chi^{(3)} E^2) E \qquad (6.94)$$

If you remember from before, the linear polarization is given by $P = \epsilon_0 \chi E$. Therefore the new susceptibility is $\chi = \chi^{(3)} E^2 = \chi^{(3)} I$, where I is the intensity. Since the susceptibility is related to the refractive index, the index itself is intensity dependent. Now a beam of light is usually thickest (i.e. most intense) in the center and then weakens at the edges so that the refractive index has the same structure. But this looks like a lens, and hence self-focusing. However, the Kerr effect can also be used for positive purposes. One is that this property can be used to mode-lock and produce very short pulses. (Can you see how this works?)

6.5   Multiphoton Processes

The non-linear phenomena as presented so far are collective effects of non-linear media which exhibit higher order non-linearities. But can each individual atom behave in a non-linear way? We have seen that in a sense each atom is a harmonic oscillator and, when the motion becomes anharmonic, the atom starts to behave in a non-linear fashion. Let us talk a bit more about the non-linear atomic behavior.

Let us now revisit the photoelectric effect using all our knowledge of non-linear optics. When explaining the photoelectric effect, Einstein wrote the energy conservation law for photons knocking out electrons from a material (i.e. a metal):

$$\frac{1}{2}mv^2 = \hbar\omega - W \tag{6.95}$$

The left hand side represents the electron's kinetic energy as it leaves the metal, while the right hand side is the difference between the photons' energy and the energy needed to knock out the electron. According to this, the photoelectric effect will therefore be observed *only* if

$$\hbar\omega > W \tag{6.96}$$

so that the lowest frequency able to achieve the effect is

$$\omega_0 = \frac{W}{\hbar} \tag{6.97}$$

This effect, i.e. the speed of exiting electrons, is nothing to do with the intensity of radiation and depends only on the frequency — this was Einstein's great success in explaining the phenomenon for the first time using Planck's quantum hypothesis of radiation. But was this ultimately correct? Could we observe the effect at lower frequencies? Before the advent of lasers the answer would definitely be "no" since there were no sources of light intense enough. But suppose that the electron absorbed N photons at the same time. Then the photoelectric equation changes to

$$\frac{1}{2}mv^2 = N\hbar\omega - W \tag{6.98}$$

as we have N photons now, and the threshold frequency is given by

$$\omega_{\text{thres}} = \frac{W}{N\hbar} \tag{6.99}$$

which is N times smaller than according to Einstein's formula. This process is possible in reality, but intensities need to be high in order for the probability to absorb N photons be high. To understand these processes properly, i.e. to be able to calculate probabilities for various optical interactions to take place, we need first to use the proper quantum mechanics and remind ourselves of perturbation theory.

Final Thought: When a lot of atoms emit at different times, the resulting waves will all have different phases, even though the oscillating frequencies may be the same. The resulting amplitude will be a superposition of all these waves which are out of phase with each other, so to speak. So what? Well, the point is that the intensity of this incoherent superposition (mixture might be a more appropriate term) is not going to be very high. We have seen how this works. To summarize, the amplitude is given by

$$U = e^{i(\omega t + \phi_1)} + e^{i(\omega t + \phi_2)} + \cdots + e^{(i\omega t + \phi_N)} \qquad (6.100)$$

$$= e^{i\omega t}(e^{\phi_1} + e^{i\phi_2} + \cdots + e^{i\phi_N}) \qquad (6.101)$$

Since the phases are truly random, for every phase ϕ there is a phase $-\phi$ in the summation. To obtain the intensity we need to take a mod square of this expression. So,

$$|U|^2 \sim N + 2\sum_{i \geq j} \cos(\phi_i - \phi_j) \qquad (6.102)$$

But the average of the summation is zero since all the phases are random! Therefore the intensity is proportional to the number of waves.

But what if we can somehow ensure that they all have the same phase? Then,

$$I = |U|^2 \sim |e^{i\omega t}|^2 |(e^\phi + e^{i\phi} + \cdots + e^{i\phi})|^2 = N^2 \qquad (6.103)$$

which is the highest intensity we can imagine from N interfering waves! It is the square of the incoherent mixture of waves. Thus if we want to see some light at some frequency properly, we had better make sure that all the different wave contributions at that frequency are in phase — hence the term "phase matching", which lies behind the phenomena of mode-locking and high harmonic generation. These two are the most important phenomena to understand from this chapter.

We now leave the world of classical optics (and physics). What we have learnt in the first six chapters is the best that can be done with classical physics and the simplistic treatments developed by Planck and Einstein (and a few other people).

Chapter 7

Interactions of Light
with Matter II

Now we will go into the details of the quantum description of the atom field interaction and explain phenomena that are beyond the purely classical treatment. A very good textbook on quantum mechanics that goes well beyond the material we cover in this book is [Diu, Laloe and Cohen-Tannoudji (1977)]. We will assume that all the problems can be treated using Schrödinger's evolution of quantized atoms where the field is still going to be treated classically.[1] Even though this is still only an approximation, it turns out to be a very good one that captures all the main features of light–matter interaction. There are, of course, shortcomings in this approach, such as the inability to derive the spontaneous emission (neither the mechanism nor the A coefficient) and to explain the effects such as Casimir's or the Lamb shift, to name a few. But on the whole, very frequently in atomic physics and quantum optics we never need to go beyond this level of approximation. This is why we will dedicate two whole chapters to it.

7.1 Vector Spaces

It turns out that in order to describe physical processes properly we need to represent states of physical systems as vectors. Furthermore, these vectors need to live in a complex vector space called a Hilbert space. (No need to worry about this name too much. You won't need to master all the mathematical technicalities in order to understand most of quantum optics; for a detailed exposition of Hilbert spaces I would recommend the pioneering book of von Neumann entitled *The Mathematical Foundations of Quantum*

---

[1] Historically, quantum theory of fields developed right after quantum mechanics was introduced in 1924. Still, it took some time to realize exactly how to quantize light. In spite of 80 years of work gravity is still not quantized and this is probably one of the biggest open problems in physics at present.

Theory [von Neumann (1955)], but it will not be necessary to understand the rest of the book.) As I explained before, the evidence is overwhelmingly in favor of quantum superpositions with complex amplitudes (remember the Mach–Zehnder interferometer at the beginning of the book and the fact that there was a phase difference i between the two ways that the photon could take through a beam splitter).[2] The vector structure itself is not so surprising since we can think of the states of classical fields (e.g. the electromagnetic field) as vectors as well and then add these up to obtain new states. However, the amplitudes in this case are always real numbers (yes, we did use complex numbers such as $e^{i\omega t}$, but this was a mathematical convenience of representing "cos" and "sin" functions and we always had to take the real part in order to obtain measurable quantities). Quantum mechanics is therefore radically different. Suppose that we only have two eigenstates, ψ_1 and ψ_2, which can be thought of as two orthogonal vectors (meaning that their inner product, to be defined more precisely, is equal to zero). Physically, orthogonality implies that the two states are completely distinguishable in principle.[3] Then any other state is a superposition of the two basis states:

$$\psi = \alpha\psi_1 + \beta\psi_2 \tag{7.1}$$

Contrast this with unit vectors $\mathbf{i}$ and $\mathbf{j}$. Any vector $\mathbf{v}$ in this two-dimensional space can be written as

$$\mathbf{v} = a\mathbf{i} + b\mathbf{j} \tag{7.2}$$

How do we find the coefficient a? We take the dot product of $\mathbf{v}$ and $\mathbf{i}$, so that

$$a = \mathbf{v}\mathbf{i} \tag{7.3}$$

and likewise $b = \mathbf{v}\mathbf{j}$. This works because $\mathbf{i}$ and $\mathbf{j}$ are orthogonal, i.e. $\mathbf{i}\mathbf{j} = 0$. In quantum mechanics states are described by (wave) functions, and functions themselves behave like vectors. We can, for example, choose an orthogonal basis of functions and express other functions in terms of this basis. The Fourier series provides

---

[2]The largest "material" object that has been measured to display diffraction (superposition) properties to date is the "buckey ball" molecule, $C60$, consisting of 60 carbon atoms. The experiment is by M. Arndt *et. al*, 680–682, 14 October 1999. Yours truly does not believe that there will be any limitations to the superposition principle.

[3]It is possible to have two different states in quantum theory which are not completely distinguishable. This, in fact, is the whole point behind the concept of a superposition.

one possible basis. In equation (7.1), we can find expansion coefficients by finding the overlap between ψ and ψ_1 and ψ and ψ_2
respectively:

$$\alpha = \int \psi(x)\psi_1(x)dx \tag{7.4}$$

$$\beta = \int \psi(x)\psi_2(x)dx \tag{7.5}$$

Again, as in the case of vectors, this assumes that the two functions
themselves are orthogonal, i.e. have no overlap:

$$\int \psi_1(x)\psi_2(x)dx = 0 \tag{7.6}$$

Now since quantum states need to be described by complex valued
functions, the overlap has to be defined as

Overlap
Integral

$$\int \psi_1^*(x)\psi_2(x)dx = 0 \tag{7.7}$$

where the "star" denotes the complex conjugate operation (and
stars are also needed in the above equations for α and β).

Now we need a way of converting these complex quantities
(quantum states) into "something real" that we can actually measure. The basic rule of quantum mechanics, called the *projection
postulate*, is the following. Suppose that the system is in a superposition of two orthogonal states ψ_1 and ψ_2:

$$\psi = a_1\psi_1 + a_2\psi_2 \tag{7.8}$$

Then the probability of obtaining ψ_1 after a measurement is $|a_1|^2$
and the state after this outcome is ψ_1. Alternatively, we can obtain
ψ_2 with the probability $|a_2|^2$. The outcome is otherwise completely
random with the given probability distribution. Note that we can
make a measurement in quantum mechanics in a different basis. For
example, we can choose to measure the basis $\psi_\pm = (\psi_1 \pm \psi_2)/\sqrt{2}$.
The state ψ can be expressed in the new basis as

$$\psi = \frac{(a_1 + a_2)}{\sqrt{2}}\psi_+ + \frac{(a_1 - a_2)}{\sqrt{2}}\psi_- \tag{7.9}$$

The probabilities are now easy to infer from the mod square rule,
and we obtain $|a_1 \pm a_2|^2/2$.

But we don't just observe probabilities (i.e. the frequencies of
occurrences of certain outcomes), we also observe positions, momenta and so on. For this we need to introduce the concept of an

operator. We will now introduce a different way of representing quantum mechanics which will be very useful to us later on.

7.2   Dirac Formalism

Quantum mechanics, as you already know, is based on four postulates. (Perhaps you have never heard it presented this way, but now it is the right time to face the truth!)   First, states of physical systems are represented by vectors in a complex vector space (known as a Hilbert space). Note that functions that you have been using to represent states so far can also be thought of as vectors. Secondly, physical quantities that are observable are represented by Hermitian operators, whose eigenvalues are real, and eigenvectors are orthonormal. Thirdly, the expectation value of an observable $\hat{O}$ in a state $\Psi(x,t)$ is given by the integral (this is the same as the above-mentioned projection postulate) **Quantum Postulates**

$$\langle \hat{O} \rangle = \int \Psi^*(x,t)\hat{O}\Psi(x,t)dx \qquad (7.10)$$

(this average is a function of time).   Finally, the evolution of a closed system is given by the Schrödinger equation

$$H(x,t)\Psi(x,t) = i\hbar\frac{\partial\Psi}{\partial t} \qquad (7.11)$$

where $H(x,t)$ is the so-called Hamiltonian operator, representing the total energy of the system we are describing. I will have much more to say about this later on and we will solve a number of problems which will acquaint us well with the formalism altogether (this is the only way of learning it anyway).

The above notation is written in the position basis (i.e. in terms of x). Now, there is nothing special about position; we may equally well use the momentum or any other observable. It is also cumbersome to write integrals and one gets very tired of it after having to use it all the time. Dirac suggested the following notation instead:

$$\langle \hat{O} \rangle = \langle\Psi|\hat{O}|\Psi\rangle \qquad (7.12)$$

In this notation the overlap between two states $\Psi(x,t)$ and $\Phi(x,t)$ is given by

$$\langle\Psi|\Phi\rangle \qquad (7.13)$$

Let's give an example to see how this correspondence works. The harmonic oscillator of mass m and natural frequency ω_c has the following eigenstates corresponding to the lowest three eigenvalues $1/2\hbar\omega_c, 3/2\hbar\omega_c$ and $5/2\hbar\omega_c$ (to be derived in detail later in the book):

$$u_0 = (m\omega_c/\pi\hbar)^{1/4}\exp(-m\omega_c x^2/\pi\hbar) \tag{7.14}$$

$$u_1 = (4/\pi)^{1/4}(m\omega_c/\hbar)^{3/4}x\exp(-m\omega_c x^2/\pi\hbar) \tag{7.15}$$

$$u_2 = (m\omega_c/4\pi\hbar)^{1/4}[2(m\omega_c/\hbar)x^2 - 1]\exp(-m\omega_c x^2/\pi\hbar) \tag{7.16}$$

We will label these states as $|0\rangle$, $|1\rangle$ and $|2\rangle$ correspondingly, which signifies that the states contain no excitations (say photons), one photon and two photons respectively (this will be discussed in much more detail later on). This notation, as we will see shortly, will greatly simplify our calculations in the coming part of the book. All professional quantum physicists use it and life would really be much more difficult without it.

Note that the orthogonality conditions between different eigenstates are expressed in a very elegant way as

$$\langle i|j\rangle = \delta_{ij} \tag{7.17}$$

so that, for example,

$$\langle 0|1\rangle = \langle 1|0\rangle = 0 \tag{7.18}$$

but also

$$\langle 0|0\rangle = \langle 1|1\rangle = 1 \tag{7.19}$$

meaning that the states are normalized (all the probabilities add up to one in other words).

And now, the crux of the formalism — the icing on the cake as it were — operators. Operators are "objects" that act on states and transform one state into another one — as simple as that! Operators represent evolutions of quantum systems and various observables that we can measure on our system under observation. Let me first give an example and then we will apply it later to more complex systems. I recall our favorite quantum example: the Mach–Zehnder interferometer (shown in Figures 1.4 and 2.1). The initial states of the photon are $|0\rangle$ and $|1\rangle$ (referring to the entrance port of the beam splitter). How does the beam splitter now act? Well, it sends $|0\rangle$ into $|2\rangle + i|3\rangle$ and sends $|1\rangle$ into $i|2\rangle + |3\rangle$ (2 and 3 are the output ports). (The imaginary phase is there because the reflection at the right angle generates this phase; can you see how to derive this?) The beam splitter executes what is known as a unitary transformation. Every quantum evolution is described by a unitary transformation — which is always reversible (apart from this transformation we also have a measurement according to the conventional, so-called "Copenhagen", rules of quantum mechanics — but we will discuss this later in more detail). The main property of any unitary transformation is that it sends orthogonal (input) states into orthogonal (output) states. Is that true for the beam splitter? The overlap of the output states is

$$(-i\langle 2| + \langle 3|)(|2\rangle + i|3\rangle) = -i + i = 0 \qquad (7.20)$$

and so they are orthogonal indeed! But what does the operator representing the beam splitter look like? The equation is

$$\hat{U}_{BS} = (|2\rangle + i|3\rangle)\langle 0| + (i|2\rangle + |3\rangle)\langle 1| \qquad (7.21)$$

where the hats as usual signify the operators. Let's see what happens when the beam splitter operates on $|0\rangle$. We get

$$\{((|2\rangle + i|3\rangle)\langle 0| + (i|2\rangle + |3\rangle)\langle 1|\}|0\rangle = (|2\rangle + i|3\rangle)\langle 0|0\rangle = (|2\rangle + i|3\rangle)$$
$$(7.22)$$

Bingo! We get the right result. Formally, we say that the unitary transformation U has two defining properties:

(1) The operator representing the unitary evolution is invertible (it has an inverse).
(2) $U^\dagger = U^{-1}$, i.e. the inverse U^{-1}, is the same as the Hermitian conjugate (i.e. conjugate transpose), $U^\dagger$, of the operator.

Let's now see Dirac's notation in some more action. Say that the Hamiltonian (of a two level system) is

$$\hat{H} = E_0|0\rangle\langle 0| + E_1|1\rangle\langle 1| \qquad (7.23)$$

This, as I briefly stated, represents the total energy of the two level system. So, the energy of the state $|0\rangle$ is E_0 and the energy of the state $|1\rangle$ is E_1. We can, of course, shift the whole energy spectrum level by any arbitrary amount, δ, and this won't change the physics. The only change will be an additional phase to the whole state of the system equal to $e^{-i\delta t/\hbar}$, which itself is not observable.

Supposing that the initial state of the system is $|0\rangle + |1\rangle$, what is the state of the system at time t if the system evolves according to $\hat{H}$? We have to solve the Schrödinger equation in order to obtain the correct form of this. So,

$$(E_0|0\rangle\langle 0| + E_1|1\rangle\langle 1|)|\psi(t)\rangle = i\hbar\frac{\partial|\psi(t)\rangle}{\partial t} \qquad (7.24)$$

Let's assume that the solution is of the (most general) form

$$a(t)|0\rangle + b(t)|1\rangle \qquad (7.25)$$

Plugging this into the evolution equation leads to

$$a(t)E_0|0\rangle + b(t)E_1|1\rangle = i\hbar(\dot{a}(t)|0\rangle + \dot{b}(t)|1\rangle) \qquad (7.26)$$

And now the standard trick: multiply from the left by $\langle 0|$ and then by $\langle 1|$. This gives us two equations:

$$\dot{a}(t) = -i\omega_0 a(t) \qquad (7.27)$$
$$\dot{b}(t) = -i\omega_1 b(t) \qquad (7.28)$$

where $\omega_i = E_i/\hbar$. These two are easily integrated to yield

$$a(t) = a(0)e^{-i\omega_0 t} \qquad (7.29)$$
$$b(t) = b(0)e^{-i\omega_1 t} \qquad (7.30)$$

Remembering that $a(0) = b(0) = 1$, the state at time t is equal to

$$|\psi(t)\rangle = e^{-i\omega_0 t}|0\rangle + e^{-i\omega_1 t}|1\rangle = e^{-i\omega_0 t}(|0\rangle + e^{-i(\omega_1 - \omega_0)t}|1\rangle) \quad (7.31)$$

and so the system "precesses" at the frequency $(\omega_0 - \omega_1)$ (again the overall phase $e^{-i\omega_0 t}$ is physically irrelevant).

Now, we have solved our first quantum optical (in fact, quantum mechanical in general) problem, and I'd like to reflect on the nature of fundamental quantities in quantum mechanics as opposed to classical physics. We saw above that the central role was played by the state of the system, and its Hamiltonian. Once the initial state is known, and we have the Hamiltonian, we can calculate anything we want at any future time of the evolution of the system. In classical physics, on the other hand, forces play the key role. Forces determine how the system will behave through Newton's laws. In quantum physics, forces are not fundamental — in fact, in some sense they don't really exist, but can only be talked of approximately. So, there is a radical shift in emphasis in our view of the world.

Note, however, that the Hamiltonian in our simple example above is time independent (i.e. E_0 and E_1 don't depend on t). This is why the above equation is very easy to solve. What happens if the Hamiltonian is time dependent?

7.3   Time Dependent Perturbation Theory

I will now introduce a method for solving the Schrödinger equation when the Hamiltonian is time dependent. This is impossible to do analytically in general and for many important applications we will have to resort to approximations. The theory developed for this purpose is the time dependent perturbation theory. It was developed by Dirac in the 1930s. This is most likely one of the most difficult pieces of mathematics to swallow when learning about quantum mechanics (optics in particular). I will first present it the way it is conventionally done and then do it in a simpler way. The simpler way is mathematically a bit less clean, but it is the way that I think about perturbations and do my calculations. I hope that after you have gone through the two formalisms, you will see there is not much to this and it will become second nature after a few calculations (just like everything else does in mathematics after a bit of practice).

Why do we need a time dependent Hamiltonian in optics? Because we'd like to study interactions of atoms with light (and not only that) and the field is usually time varying. Light, as we know, is an electromagnetic wave and is described by six numbers at each position in space changing in time (three electric and three magnetic components). Therefore, the resulting Hamiltonian will be time dependent. However, light will in this chapter still be treated as a bunch of numbers (and NOT operators) oscillating in time. This is basically a classical description (quantum mechanically these will have to be upgraded to, in general, non-commuting operators). The picture where atoms are quantized, but light is still kept classical (and enters only through numbers in the Hamiltonian) is called the semi-classical approximation. The Schrödinger equation in the semi-classical approximation is given in the following way:

$$i\hbar \frac{\partial}{\partial t}|\Psi(t)\rangle = (\hat{H}_0 + \hat{V}(t))|\Psi(t)\rangle \qquad (7.32)$$

Here we have assumed that there is a free Hamiltonian which is time independent and all the time dependence is put into the interaction potential $\hat{V}(t)$. Why is this equation difficult to solve in general, as I claimed before? Why isn't the solution just obtained by exponentiation of the right hand side so that we have

$$|\Psi(t)\rangle = e^{-i\int(\hat{H}_0 + \hat{V}(t))dt}|\Psi(0)\rangle \qquad (7.33)$$

This solution is correct only if the interaction potentials at different times commute, i.e. if $\hat{V}(t)\hat{V}(t') = \hat{V}(t')\hat{V}(t)$ (for all times t, t'). The reason for this is that, in general, it is not true that

$$e^{\hat{A}}e^{\hat{B}} = e^{\hat{A}+\hat{B}} \qquad (7.34)$$

where A and B are two operators. The above is true only if the two operators commute $\hat{A}\hat{B} = \hat{B}\hat{A}$. Otherwise, we have to employ the perturbation series method to solve the time dependent Schrödinger equation. Here is how this goes.

We first perform a standard operation in order to eliminate the non-perturbed part of the Hamiltonian: this procedure is known as "going into the interaction picture". This we do to make our life easier, since we already know how the system evolves under a time independent Hamiltonian. We first define a new wave function

$$|\Psi(t)\rangle = e^{iH_0 t/\hbar}|\psi(t)\rangle \qquad (7.35)$$

which leads to a new Schrödinger equation, now in the interaction picture

Interaction Picture

$$i\hbar\frac{\partial}{\partial t}\psi(t) = \tilde{V}(t)\psi(t) \qquad (7.36)$$

with the modified interaction Hamiltonian

$$\tilde{V}(t) = e^{i\hat{H}_0 t/\hbar}\hat{V}(t)e^{-iH_0 t/\hbar} \qquad (7.37)$$

A solution to this equation can formally be written as

$$|\psi_i(t)\rangle = |i\rangle + \left(-\frac{i}{\hbar}\right)\int dt'\tilde{V}(t')|\psi_i(t')\rangle \qquad (7.38)$$

What we would like to compute is the amplitude to go from the initial to the final state $\langle f|\hat{A}(t)|i\rangle$. This amplitude changes with time and is defined via

$$|\psi_i(t)\rangle = \hat{U}(t)|i\rangle \qquad (7.39)$$

where, as discussed before, U is a unitary operator. Any time we solve the Schrödinger equation, thereby "exponentiating the Hamiltonian", we always obtain a unitary operator (show in one line that the Hermitian conjugate of $e^{i\hat{H}}$, where $\hat{H}$ is Hermitian is, in fact, equal to its inverse, i.e. $e^{-i\hat{H}}$). Thus,

$$\hat{U}(t) = 1 + \left(-\frac{i}{\hbar}\right)\int_{t_0}^{t} dt'\hat{V}(t')\hat{U}(t') \qquad (7.40)$$

The operator describing the evolution of a quantum state is called a unitary operator. Unitary operators are invertible as well satisfying the rule that $\hat{U}\hat{U}^\dagger = \hat{U}^\dagger\hat{U} = I$. We have seen that in order to obtain the unitary evolution operator we have had to (in some sense) exponentiate the Hamiltonian H. We can think of Hermitian operators as generalizations of real numbers (their eigenvalues have to be real), in which case the unitary operators are generalizations of unit modulus complex numbers of the form e^{ir}, where r is real.

The standard way of solving the evolution equation is by "an iterated substitution of this equation into itself", a method leading to the so-called Dyson series **Dyson Series**

$$\hat{U}(t) = 1 + \sum \hat{U}^{(n)}(t) \qquad (7.41)$$

where

$$\hat{U}^{(1)} = \left(-\frac{i}{\hbar}\right) \int_{t_0}^{t} dt_1 \tilde{V}(t_1) \qquad (7.42)$$

$$\hat{U}^{(2)} = \left(-\frac{i}{\hbar}\right)^2 \int_{t_0}^{t} dt_1 \tilde{V}(t_1) \int_{t_0}^{t_1} dt_2 \tilde{V}(t_2) \qquad (7.43)$$

$$\hat{U}^{(n)} = \left(-\frac{i}{\hbar}\right)^n \int_{t_0}^{t} dt_1 \tilde{V}(t_1) \int_{t_0}^{t_1} dt_2 \tilde{V}(t_2) \cdots \int_{t_0}^{t_{n-1}} dt_n \tilde{V}(t_n) \cdots \qquad (7.44)$$

The physical interpretation of this series is very intriguing (see Figure 7.1). We can go from the initial state to the final state without any application of the perturbation, or with one application of perturbation, or two applications and so on. In most of our applications we will only be interested in the first order approximation which will have the highest probability.

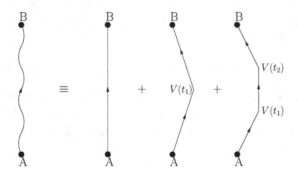

Fig. 7.1 The Dyson perturbation expansion. Each term contributes to the final amplitude for a given process in quantum mechanics. The first term is the zeroth order where nothing happens, then we have the first order with one Hamiltonian application and so on.

Let's give an example to get a clearer picture: we revisit multiphoton processes. When an atom goes from one state to another it can do so by absorbing one photon, which is basically represented by one application of $\tilde{V}$. However, it could also do so by absorbing two photons (i.e. two applications of $\tilde{V}^2$); the atom would go from i to some intermediate state m via $\tilde{V}$ and then from m to f via another $\tilde{V}$. This is written as (the reading is from right to left like in the Arabic language): $\langle f|\tilde{V}|m\rangle\langle m|\tilde{V}|i\rangle$. Now each of these two processes is roughly as likely as one photon process. The probability of both processes is the probability of one followed by the other, which is why the probability for two photon process (second order process) is the square of the single photon process (first order process). This is exactly why most of the time we can ignore higher order processes: suppose that the probability for a single photon absorption is 10^{-2}, then two photon process is only about 10^{-4} likely. It also explains why we need high intensities to observe higher order processes as mentioned before.

I'd now like to apply this briefly to emission and absorption of light to show you how it works in the semi-classical picture. We will do this in more detail in the next section, but I don't want to keep you in suspense for too long. We have to work out what $\hat{V}$ is when atoms interact with light. An atom is a small electric dipole D (the electron and the nucleus are the negative and the positive charges respectively, separated by some distance depending on the atom). Classically the energy of a dipole **D**, in an external electric field **E**, is given by the dot product of the dipole and the field, $V = \mathbf{DE}$. We have to quantize this, but we remember that we don't quantize the field at this stage. Therefore the field–dipole interaction is given by

$$\hat{V}(t) = \hat{D}E(t) \qquad (7.45)$$

Field–Dipole Interaction

where $\hat{D}$ is the quantized electric dipole. Note that because the magnetic field is much smaller in magnitude we ignore its effect completely in this case.[4] The electric field is assumed to be oscillating at frequency ω and is given by

$$E(t) = \frac{E}{2}e^{-i\omega t} + \frac{E^*}{2}e^{i\omega t} \qquad (7.46)$$

The first order correction from the time dependent Dyson expansion now has the form

$$\langle f|\hat{U}^{(1)}(t)|i\rangle = (-i/\hbar)\left(\int dt_1 e^{-i(\omega_i - \omega_f - \omega)t}\frac{E}{2}\langle f|\hat{D}|i\rangle \right. \qquad (7.47)$$

$$\left. + \int dt_1 e^{-i(\omega_i - \omega_f + \omega)t}\frac{E^*}{2}\langle f|\hat{D}|i\rangle\right) \qquad (7.48)$$

---

[4]Later on we will consider a neutron spin interacting with a magnetic field and the electric field will be neglected.

The factors containing ω_i and ω_f come from the "free" evolving Hamiltonian $\hat{H}_0$ as in equation (7.31). Let's assume that the interaction is much longer than the characteristic atomic periods and take the limit of infinite time (this is, strictly speaking, incorrect to do as then the first order perturbation would be invalid). We obtain

$$\langle f|\hat{U}^{(1)}(t)|i\rangle = (-2\pi i/\hbar)\left(\delta(\omega_i - \omega_f - \omega)\frac{E}{2}\langle f|\hat{D}|i\rangle \right. \quad (7.49)$$

$$\left. + \delta(\omega_i - \omega_f + \omega)\frac{E^*}{2}\langle f|\hat{D}|i\rangle\right) \quad (7.50)$$

(this result follows from the property of Fourier transforms). Therefore we see two terms corresponding to the emission ($\omega = \omega_f - \omega_i$) and absorption ($\omega = \omega_i - \omega_f$) of light. Note that the frequency of light in this limit has to be exactly equal to the transition frequency just like in both the classical and Einstein's model.

7.4 Alternative Derivation of Perturbation

Instead of looking at the Schrödinger equation as describing the evolution of the quantum state we can look at it as the evolution of the unitary operator acting on the state. I would like to derive the time dependent perturbation treatment in this way as it makes it easier for us to understand what is really going on there. From Schrödinger's equation we have that

$$i\hbar\frac{\partial \hat{U}(t,t_0)}{\partial t} = \hat{V}(t)\hat{U}(t,t_0) \quad (7.51)$$

where $\hat{U}(t,t_0)$ is the (unitary) evolution operator describing the evolution of the initial state so that

$$|\psi(t)\rangle = \hat{U}(t,t_0)|\psi(t_0)\rangle \quad (7.52)$$

The above equation can be derived differentiating this expression. We obtain

$$\frac{\partial|\psi(t)\rangle}{\partial t} = \frac{\partial \hat{U}(t,t_0)}{\partial t}|\psi(t_0)\rangle \quad (7.53)$$

Multiplying both sides by $i\hbar$ and using Schrödinger's equation we see that

$$\hat{H}\hat{U}|\psi(t_0)\rangle = i\hbar\frac{\partial \hat{U}(t,t_0)}{\partial t}|\psi(t_0)\rangle \quad (7.54)$$

This has to be true for all $|\psi(t_0)\rangle$ and so we arrive at the desired equation.

Let's discretize time by introducing a small time step

$$\Delta t = \frac{t - t_0}{N} \qquad (7.55)$$

Now, the evolution equation can be written for a short time period in the form (I am now absorbing $\hbar$ into V just for shorthand purposes)

$$U(t_0 + \Delta t, t_0) = [1 - iV(t_0)\Delta t] \qquad (7.56)$$

and, since $U(t, t_0) = U(t, t_{N-1})U(t_{N-1}, t_{N-2}) \cdots U(t_1, t_0)$, we have that

$$U(t, t_0) = [1 - iV(t_N)\Delta t][1 - iV(t_{N-1})\Delta t] \ldots [1 - iV(t_0)\Delta t]$$

$$= 1 - i \sum_n V(t_n) + (-i)^2 \sum_{n<m} V(t_n)V(t_m) \qquad (7.57)$$

$$+ (-i)^3 \sum_{n<m<l} V(t_n)V(t_m)V(t_l) + \cdots \qquad (7.58)$$

This relationship is now the same as the Dyson series we obtained before, if we take the limit of $\Delta t \to 0$ and all the sums become integrals. We notice that the final (total) value of the evolution operator is made up of series of "perturbations" as before. Note that they obey the "typical" time ordering convention: perturbations are applied one after another.

Evolution under time dependent Hamiltonians can be handled by breaking the evolution into a large number of small time steps. In each of these time steps we can assume that the Hamiltonian is constant. So we would have something like $e^{-iH(t_N-t_{N-1}))\delta t} \ldots e^{-iH(t_3-t_2))\delta t} e^{-iH(t_2-t_1))\delta t}$. Since δt is small this can be written as $(1 - iH(t_N - t_{N-1})\delta t) \cdots (1 - iH(3 - t_2)\delta t)(1 - iH(t_2 - t_1)\delta t)$. This can be expanded to yield $1 + i \sum_i H(t_i - t_{i-1})\delta t +$ higher order terms, which is exactly the result of perturbation we derived previously. So, the most general question we can ever ask in quantum mechanics, in fact in physics, is: if we start in the state $|i\rangle$, what is the probability to end up in the state $|f\rangle$ after the time T long evolution via the Hamiltonian H? The answer is given by the mod square of the amplitude

$$a_{if} = \langle f|(1 - iH(t_N - t_{N-1})\delta t) \cdots (1 - iH(t_2 - t_1)\delta t)|i\rangle \qquad (7.59)$$

$$= \langle f| \left(1 + i\sum_i H(t_i - t_{i-1})\delta t + \text{higher order terms}\right)|f\rangle \qquad (7.60)$$

(Note that H should be divided through by $\hbar$ everywhere). Let us now describe one important application of perturbation theory.

7.5   The Wigner–Weisskopf Theory

Now we will be for the first time in the position to derive stimulated emission (absorption) rigorously and obtain the correct expression for the Einstein B coefficient. If you remember the beginning of the book, this was something that Einstein himself was unable to derive as he didn't know enough quantum mechanics. We now know as much as anyone did in 1930. (Note, though, that we still have some 70 years to learn!) This is possibly the most beautiful part of the second third of the course, both in terms of its elegance and its wide range of applicability. Among other things we will be showing how to prove the second law of thermodynamics (you probably never thought you'd find something like that in a quantum optics textbook!). So, let's go for it.

7.5.1   *Constant perturbation*

First, let us do the easy thing: constant potential, i.e.

$$\hat{V}(t) = V \tag{7.61}$$

Suppose that at $t = 0$ the system is in the state $|i\rangle$. Then

$$c_n^{(1)} = \frac{-i}{\hbar} V_{ni} \int e^{i\omega_{ni}t'} dt' \tag{7.62}$$

$$= \frac{V_{ni}}{E_n - E_i}(1 - e^{i\omega_{ni}t}) \tag{7.63}$$

where $V_{ni} = \langle n|\hat{V}|i\rangle$ is the transition element between the levels $|i\rangle$ and $|n\rangle$. And, so the probability for a transition to the nth level from the initial state is given by

$$|c_n^{(1)}|^2 = \frac{4|V_{ni}|^2}{|E_n - E_i|^2} \sin^2 \frac{(E_n - E_i)t}{2\hbar} \tag{7.64}$$

Here I have used the fact that $1 - e^{i\theta} = e^{i\theta/2}(e^{-i\theta/2} - e^{i\theta/2})$, so that $|1 - e^{i\theta}|^2 = 2\sin^2(\theta/2)$. For (sufficiently) short times we have

$$|c_n^{(1)}|^2 = \frac{1}{\hbar^2}|V_{ni}|^2t^2 \tag{7.65}$$

This is a bit disturbing — the dependence is quadratic. We said that the idea of this theory is to derive the stimulated emission, but this is an exponential process as we saw at the very beginning. The decay is of the form $e^{-\Gamma t}$, which is for short times $1 - \Gamma t$, i.e. linear in t! So, we seem to have a problem. However, and here comes in Dirac's smart idea (generalized by Wigner and Weisskopf). He reasoned that in general many different frequencies will interact with the system and this will happen incoherently. In the end the total probability will be an average over all different probabilities for different frequencies. This leads to

$$\sum_n |c_n^{(1)}|^2 = \int dE_n \rho(E_n) |c_n^{(1)}|^2 \tag{7.66}$$

$$= 4 \int \frac{|V_{ni}|^2}{|E_n - E_i|^2} \sin^2 \frac{(E_n - E_i)t}{2\hbar} \rho(E_n) dE_n \tag{7.67}$$

As time becomes larger and larger, we can use the following result

$$\lim_{t \to \infty} \frac{1}{|E_n - E_i|^2} \sin^2 \frac{(E_n - E_i)t}{2\hbar} = \frac{\pi t}{2\hbar} \delta(E_n - E_i) \tag{7.68}$$

Therefore,

$$\sum_n |c_n^{(1)}|^2 \to \frac{2\pi}{\hbar} |V_{ni}|^2 \rho(E_n) t|_{E_n = E_i} \tag{7.69}$$

Note the comforting appearance of the linear time dependence.

What about the transition rate between the levels? It is given by the following expression

$$w_{i \to n} = \frac{2\pi}{\hbar} |V_{ni}|^2 \rho(E_n) \delta(E_n - E_i) \tag{7.70}$$

Fermi's Golden Rule

This formula is known as Fermi's golden rule as it is of huge importance in quantum mechanics and, in particular, quantum optics. It was derived by Dirac in late 1920s and it's not the only formula in physics that is not named after its inventor![5]

Now we consider a perturbation that is the cornerstone of quantum optics (and not only quantum optics — it is an archetypical calculation whenever we have any kind of field interacting with matter). The following section is an extremely important piece of physics.

7.5.2 *Harmonic perturbation*

The harmonic perturbation is given by

$$V(t) = V e^{i\omega t} + V^* e^{-i\omega t} \tag{7.71}$$

This is just the usual form of the "oscillating" electric (or magnetic, or any other) field. Let's "turn on" the first order perturbation

---

[5] According to the mathematician Arnold most inventions are not named after the original inventor.

formula and obtain

$$c_n^{(1)} = \frac{-i}{\hbar} \int (V_{ni}e^{i\omega t'} + V_{ni}^* e^{-i\omega t'})e^{i\omega_{ni}t'}\,dt' \qquad (7.72)$$

$$= \frac{1}{\hbar}\left[\frac{1 - e^{i(\omega + \omega_{ni})t}}{\omega + \omega_{ni}}V_{ni} + \frac{1 - e^{i(\omega - \omega_{ni})t}}{\omega - \omega_{ni}}V_{ni}^*\right] \qquad (7.73)$$

This is similar to the constant perturbation case. So as t gets larger, the only noticeable effect happens if $E_n = E_i - \hbar\omega$ (stimulated emission) or if $E_n = E_i + \hbar\omega$ (stimulated absorption).

And here is an amazing thing. In complete analogy with the constant perturbation case we have

$$w_{i\to n} = \frac{2\pi}{\hbar}|V_{ni}|^2\delta(E_n - E_i \pm \hbar\omega) \qquad (7.74)$$

But,

$$|V_{ni}|^2 = |V_{in}|^2 \qquad (7.75)$$

since $\langle i|V|n\rangle = \langle n|V|i\rangle^*$, and so their mod squares are equal (we remember that V is a Hermitian operator so $V^\dagger = V$). So we have now rigorously proved that the stimulated emission and absorption rates are equal to each other (as derived by Einstein in a much more heuristic way).

So we have the basis for stimulated absorption and stimulated emission. We will use this later on to derive Einstein's B coefficient. Thus we can also derive the form of the A coefficient since the A and the B coefficient are mutually related. But what about deriving the spontaneous emission rigorously from the above perturbative analysis? It turns out that we can't really derive it in the semi-classical picture! This is because it is meaningless to have no perturbation and still have spontaneous emission — there is nowhere in the semi-classical picture it could possibly come from. So we need the full quantum mechanical treatment with the field quantized as well. Now you see how we are starting to build up to the grand finale as we slowly enter the last part of our treatment — the "fully quantized quantum optics". But, before that, we have as promised a short digression: a derivation of the second law of thermodynamics from quantum mechanics!

Final Thought: The formalism of time dependent perturbation theory is probably one of the most important pieces of physics you will learn in your undergraduate degree. It has important implications in quantum optics as we will see later on and it is indispensable in understanding how light and matter interact. However, its applicability is much wider than that and covers almost all areas of modern physics. It is used in particle physics to describe collisions and we will see its application in statistical mechanics next. In general, whenever we have interaction between two or more quantum systems (be it particles or fields — the distinction that disappears in quantum field theory) we have to describe this using an interaction Hamiltonian; the subsequent evolution is usually (almost always) impossible (and frequently unnecessary) to solve exactly. Fermi's golden rule is then employed and has been astoundingly successful in the last 70 years or so in predicting various effects.

7.6   Digression: Entropy and the Second Law

Now I'd like to address the problem that was the main driving motivation behind Planck's work. (He derived the quantum hypothesis and the blackbody radiation en route, but this wasn't the main thing!) Planck wanted to prove that in the system that consists of atoms and radiation interacting with each other, the entropy always increases. In other words, he wanted to prove the second law of thermodynamics for this case. But he never succeeded. He didn't have the full quantum theory after all, and in particular Fermi's golden rule. We are in a much better position now. I will present the modern "proof" of the second law — it's the best one we have so far, yet not completely satisfactory (hence the inverted commas in the word "proof")! The bottom line is that there is no way of deriving something irreversible, like entropy increase, from something reversible, like the Schrödinger equation. Here is why.[6]

Suppose that you prove to me that Schrödinger's equation leads to entropy increase, i.e. we start from a state of low entropy and then, after some time, we get a state of high entropy. I can then always say that if I reverse the evolution (and this is physically allowed as the transformation $t \to -t$ leads to a possible evolution), the system would go back to its original state, i.e. it would evolve from a high entropy state to a low entropy state. And this contradicts what we are trying to prove! From this it follows that a reversible evolution can only conserve the entropy, which is the only thing we can really say after all. This, of course, is not very satisfactory as everyday experience tells us that entropy and disorder

---

[6]I will here present something that is known as "Loschmidt's objection". Loschmidt used this argument against Boltzmann's derivation of entropy increase from Newton's dynamics, but the argument applies to any reversible dynamics.

increase as time goes by. This is why I'd like to give you our best shot at resolving this paradoxical situation at the heart of physics.

The second law says that entropy of an isolated system never decreases. In statistical mechanics an isolated system can be modelled as a *Markov chain*. Random variables X, Y and Z form a Markov chain (denoted by $X \to Y \to Z$) if the conditional distribution of Z depends only on Y and is conditionally independent of X. The joint probability distribution can then be written as

$$P(x, y, z) = P(x)P(y|x)P(z|y) \qquad (7.76)$$

where $P(1|2)$ is the conditional probability of observing event 1 given that event 2 has taken place. This is a standard formula from probability theory. A system with the above properties is said to have no memory since Z depends on Y only and is independent of any previous history (on X in this case). Suppose that our system now has states $1, 2, \ldots$ each one occupied with the initial probabilities $P_1, P_2, \ldots$ respectively. The evolution of this physical system is now represented by transition probabilities of the form $P(n|m)$ that the system in the state m will "jump" to the state n (as in Figure 7.2). The rate of change of a particular probability is given by

Fermi's Master Equation

$$\frac{dP_m}{dt} = \sum_n \{P(m|n)P_n - P(n|m)P_m\} \qquad (7.77)$$

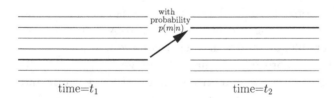

$$\text{time} = t_1 \qquad\qquad \text{time} = t_2$$

Fig. 7.2   Model of a (quantum) stochastic processes. A system can jump from any level n to any other level m as explained in the text. The jump probabilities can be calculated using perturbation theory, in particular Fermi's golden rule.

This equation is called Fermi's master equation.[7] The master equation is usually taken to be the chief statistical starting assumptions. (Note that this was not the usual starting point in

---

[7]This is a generalization of Einstein's rate equations for a two level atom interacting with radiation. It is, however, of a much more general and wider applicability and can, in fact, be taken as a starting point of thermodynamics. Note that this evolution is in general not represented by a unitary transformation. It is irreversible, unlike unitary transformations.

statistical mechanics: you would usually start by assuming Gibbs' postulates, which are only suitable for equilibrium thermodynamics. On the other hand, the master equation describes an approach to equilibrium which will not be the focus of our study.) Its interpretation is simple: the probability of the mth state changing by the system jumping from other levels to this level (the first term on the right hand side) and by the system jumping to other levels from mth level (the second term). We insert the minus sign is because the system is jumping to other levels and this reduces the probability. Since the underlying physics is reversible it turns out that $P(m|n) = P(n|m)$. This fact is referred to as the principle of *jump rate symmetry* and it is a direct consequence of Fermi's golden rule. It is worth noting that although the content of the master equation is simple and the equation seems natural, it is by no means easy to derive. In spite of this complication it has been extremely successful in thermodynamics and kinetic theory, giving complete agreement with experiment. This is why it is generally accepted as a simple and correct description of kinetics. However, it should be borne in mind that its justification is very difficult from the Schrödinger equation which is completely reversibly and entropy conserving. The master equation, on the other hand, leads to irreversible processes and we are now ready to use it to prove that the entropy of this system

Entropy

$$S = -k \sum_i P_i \ln P_i \qquad (7.78)$$

never decreases (k is Boltzmann's constant). Let us show in a few steps that this is indeed true. We have that

$$\frac{dS}{dt} = -k \sum_i \left\{ \frac{dP_i}{dt} + \frac{dP_i}{dt} \ln P_i \right\} = -k \sum_i \frac{dP_i}{dt} \ln P_i \qquad (7.79)$$

since $\sum P_i = 1$ and thus $d(\sum P_i) = 0$. Then, using the master equation with the jump rate symmetry included,

$$\frac{dS}{dt} = -k \sum_{mn} P(n|m)\{P_n - P_m\} \ln P_m \qquad (7.80)$$

$$= -\frac{k}{2} \sum_{mn} P(n|m)\{P_n - P_m\}\{\ln P_m - \ln P_n\} \qquad (7.81)$$

But each term in the sum is negative because the sign of $\{P_n - P_m\}$ is opposite to the sign of $\{\ln P_m - \ln P_n\}$. Hence this proves that

Second Law

$$\frac{dS}{dt} \geq 0 \qquad (7.82)$$

and the entropy never decreases. It should be remembered that this result relies on the master equation being a good description of the system's kinematics. Failure of this assumption could lead to entropy decrease in contradiction with the second law of thermo-dynamics. In spite of the fact that the master equation is difficult to derive rigorously in general, it is still very satisfying that the second law can be at least partly reduced to underlying (quantum) dynamics. I want to stress again that this is a part of the study of approach to equilibrium, and is something you would never do in your standard, run-of-the-mill thermodynamics course where one only studies systems already in thermal equilibrium (described by the Gibbs' postulates).

7.7   Einstein's B Coefficient

As promised earlier, we now derive Einstein's B coefficient for stim-ulated emission and absorption of radiation by a two level atom. We can derive this directly from Fermi's golden rule. We just have to assume a form of interaction between the atom and the field. This is best represented by a dipole interaction, where the Hamil-tonian is given by

$$\hat{V}(t) = e\hat{\mathbf{r}}\mathbf{E}_0 \cos \omega t \qquad (7.83)$$

The B coefficient can now be deduced from Einstein's rate equation

$$Bu(\omega) = B\rho(\omega)\langle E \rangle \qquad (7.84)$$

where u is the energy density, and ρ the density of modes of radi-ation interacting with the atom. From Fermi's golden rule, on the other hand, we know that

$$w = \frac{2\pi}{\hbar}e^2|X_{ab}|^2 E^2 \cos^2(\omega t)\frac{\rho(\omega)}{\hbar} \qquad (7.85)$$

(the extra $\hbar$ comes from the fact that $\rho(E)dE = \rho(\omega)d\omega$, so that $\rho(E) = \rho(\omega)d\omega/dE = \rho(\omega)/\hbar$). Thus we can infer that the B coefficient is

$$B = \frac{2\pi e^2|X_{ba}|^2}{\hbar^2}\frac{E_0^2}{\langle E \rangle} \qquad (7.86)$$

where $|X_{ba}| = |\langle b|X|a\rangle|$ is the dipole matrix element in the x di-rection and $|a\rangle$ and $|b\rangle$ are the two levels. If we want to express this as a function of the general dipole matrix element we need to average over all three spatial directions. Thus,

$$|X_{ba}|^2 = \frac{1}{3}|\mathbf{r}_{ab}|^2 \langle \cos^2 \theta \rangle = \frac{1}{6}|\mathbf{r}_{ab}|^2 \tag{7.87}$$

So the final expression is (assuming that the average energy is $\langle E \rangle = \epsilon_0 E_0^2$)

$$B = \frac{\pi e^2 |\mathbf{r}_{ba}|^2}{3\epsilon_0 \hbar^2} \tag{7.88}$$

B Coefficient

This is the most rigorous derivation of the B coefficient we have. Now that we know this, we can derive the A coefficient using the relationship between the A and B coefficient derived by Einstein:

$$A = \frac{\hbar \omega_{ba}^3}{\pi^2 c^3} B = \frac{e^2 \omega_{ba}^3 |\mathbf{r}_{ba}|^2}{3\epsilon_0 \hbar \pi c^3} \tag{7.89}$$

A Coefficient

But remember that this is not a rigorous derivation from the first principles of the A coefficient. This cannot be done using the semi-classical picture of the atom and field and we have to wait until we quantize light. Thus, to summarize, incoherent excitations lead to stimulated emission and absorption. However, what happens if we have coherent driving field? Suppose that the two level atom of ours could interact with one frequency only — what kind of resulting dynamics would this be? This is a good model for a highly monochromatic laser interacting with a two level atom (remember that the laser light is highly monochromatic) and we will analyze it in Chapter 8.

To close this section, let us make a link between this theory and the "classical derivation of spontaneous emission", i.e. the radiation damping analysis (Section 5.5.2). Assume that the dipole transition element $|\mathbf{r}_{ba}|^2$ is of the size of Bohr's radius:

Link with Classical Damping

$$r_B = \frac{\hbar^2}{me^2} \tag{7.90}$$

Substituting this into the A coefficient we obtain

$$A_{\text{Bohr}} = \frac{\hbar \omega^3}{3\epsilon_0 \pi c^3 m^2 e^2} \tag{7.91}$$

The subscript "Bohr" refers to the fact that the A coefficient is calculated for Bohr's orbit. The bunch of constants in the numerator and denominator remind us of the ground state energy in Bohr's model — which is, in fact, known as the Rydberg constant, and is given by

$$R = \frac{me^4}{2\hbar^2} \tag{7.92}$$

Inserting this into the formula gives

$$A_{\text{Bohr}} = \frac{e^2 \omega^3}{6\epsilon_0 \pi c^3 m(\frac{R}{\hbar})} \tag{7.93}$$

But $R/\hbar$ is the frequency of Bohr's ground state orbit. This cancels the frequency in the numerator leading to

$$A_{\text{Bohr}} = \frac{e^2 \omega^2}{6\epsilon_0 \pi c^3 m} \tag{7.94}$$

This is the same as the classical damping rate in equation (5.53)! So we see that the classical theory is pretty accurate when it comes to predicting Bohr's ground state (and hence hydrogen's) spontaneous emission rate, but fails to predict any other more complex features. Still it is impressive to see how far we can go with theory that is based on something that is about two hundred years old.

7.8 Multiphoton Processes Revisited

I'd like to show you briefly how multiphoton processes come naturally out of the formalism of time dependent perturbation theory. Suppose that we have a three level system, with levels 1 ,2 and 3, and that the frequency of the driving light field ω is such that

$$2\omega = \omega_{13} = \frac{E_3 - E_1}{\hbar} \tag{7.95}$$

where E_1 and E_3 are the energies of levels 1 and 3 respectively. Suppose that light is not "tuned" to any of the existing transitions between any two levels, i.e. $\omega \neq \omega_{nm}$, where $n = 1, 2, 3$ and $m = 1, 2, 3$. Suppose that we'd like to start from the population all in level 1 ($c_1(t = 0) = 1$) and that we'd like to raise it to level 3 by shining light of the described frequency. Let's, in addition, require that there is no coupling (Hamiltonian element) between levels 1 and 3, so that $V_{13} = V_{31} = 0$. However, suppose that coupling between 1 and 2 and 2 and 3 exists, and so $V_{12} \neq 0 \neq V_{32}$. At first sight it looks as if there is no way to instigate transitions between 1 and 3 and this conclusion is certainly right if we look only at the first order expansion in the Dyson perturbation series (check for yourself that the amplitude for this is zero). But what about the second order term? We have that

$$c_3^{(2)} = \left(-\frac{i}{\hbar}\right)^2 \int V_{32}(t_1)dt_1 \int V_{12}(t_2)dt_2 \tag{7.96}$$

and this is non-zero as both of the integrals are non-zero! In fact, if the interaction Hamiltonian is of the "dipole" form we are used to from earlier, i.e.

$$V_{nm} = d_{nm} E e^{-i\omega t} + \text{c.c.} \tag{7.97}$$

then the amplitude for the second order process is (prove it,[8] but don't forget to include the factor $e^{-i\omega_{12}t}$ and apply the short time approximation so that $e^{-i\omega t} \approx 1 - i\omega t$)

$$c_3^{(2)} = \frac{2}{\hbar^2} \frac{d_{32} d_{21} E^2}{E_2 - E_1 - \hbar\omega} t \tag{7.99}$$

(This in addition includes the fact that the terms containing the sum of the frequencies $\omega + \omega_{12}$ are ignored — which is known as the rotating wave approximation. We will talk more about this later.) Notice that the denominator involves levels 1 and 2 and not 1 and 3. How can this be explained? I leave it up to you. (Hint: what we need is $E_3 - E_1 = 2\hbar\omega$.) The conclusion is that even though we cannot go from $1 \to 3$ directly, we can use an intermediate level to aid us in this, so that we have $1 \to 2 \to 3$.

---

[8]Hint: we basically have to solve two integrals. We have

$$c_3^{(2)} \propto \int_0^{t_2} e^{-i(\omega_{12}-\omega)t_1} dt_1 \int_0^{t_2} e^{-i(\omega_{12}-\omega)t_2} dt_2 \tag{7.98}$$

Chapter 8

Two Level Systems

We now specialize to two level systems only. These are very important in quantum optics in general, since, as we have seen, it is frequently really only necessary to take two levels into account to be able to understand a great deal of phenomena. More importantly, any multilevel system can be thought of as a collection of two levels, so that our formalism generalizes immediately.

8.1 Operator Matrix Algebra

Operators are in general mappings from states into states. They transform one possible state into another one. In quantum mechanics, Hermitian operators represent observables, but they too transform states into states. So, in general we have something like

$$O|\psi\rangle = |\phi\rangle \tag{8.1}$$

which is a typical operator equation (I am deliberately omitting "hats" here as it will be obvious which quantities are operators and which are not). Now we have a nice way of using Dirac's notation (this is the beauty of Dirac's idea to represent this operator). It is

$$O = |\phi\rangle\langle\psi| \tag{8.2}$$

and so, since $\langle\psi|\psi\rangle = 1$, we have that

$$\{|\phi\rangle\langle\psi|\}|\psi\rangle = |\phi\rangle\langle\psi|\psi\rangle = |\phi\rangle \tag{8.3}$$

which is what we wanted in the first place. What if we want O to do more than that? How about finding an operator U that transforms two orthogonal states $|\psi_1\rangle$ and $|\psi_2\rangle$ (two states are orthogonal when $\langle\psi_1|\psi_2\rangle = \langle\psi_2|\psi_1\rangle = 0$) according to the following prescription

$$U|\psi_1\rangle = |\phi_1\rangle \tag{8.4}$$

$$U|\psi_2\rangle = |\phi_2\rangle \tag{8.5}$$

U can be found by writing

$$U = |\phi_1\rangle\langle\psi_1| + |\phi_2\rangle\langle\psi_2| \tag{8.6}$$

You can check for yourself that this indeed does what we need. We will use this operator notation extensively. Sometime it will, however, be more convenient to represent operators as matrices. We have the following representation hierarchy:

- Transformation $\rightarrow$ Operators $\rightarrow$ Matrices
- States $\rightarrow$ Functions $\rightarrow$ Vectors (Kets)

Let's first "upgrade" (or downgrade if you prefer; after all we are just approximating reality, aren't we?) states to vectors. We choose

$$|\psi_1\rangle = \begin{pmatrix} 0 \\ 1 \end{pmatrix} \quad |\psi_2\rangle = \begin{pmatrix} 1 \\ 0 \end{pmatrix} \tag{8.7}$$

We can equally well choose any other representation as long as the resulting vectors are orthogonal. For example,

$$|\psi_1\rangle = \frac{1}{\sqrt{2}}\begin{pmatrix} 1 \\ 1 \end{pmatrix} \quad |\psi_2\rangle = \frac{1}{\sqrt{2}}\begin{pmatrix} 1 \\ -1 \end{pmatrix} \tag{8.8}$$

is just as good (check that these two vectors are orthogonal). Now we have the operator: it has to transform a 2-vector into another 2-vector. Thus it has to be represented by a two by two matrix. The elements are

$$[U] = \begin{pmatrix} \langle\psi_1|U|\psi_1\rangle & \langle\psi_1|U|\psi_2\rangle \\ \langle\psi_2|U|\psi_1\rangle & \langle\psi_2|U|\psi_2\rangle \end{pmatrix} \tag{8.9}$$

(I am being a bit pedantic with my notation where the square brackets indicate the matrix rather than the "butterfly" form of the operator.) Substituting the form of U we obtain

$$[U] = \begin{pmatrix} \langle\psi_1|\phi_1\rangle & \langle\psi_1|\phi_2\rangle \\ \langle\psi_2|\phi_1\rangle & \langle\psi_2|\phi_2\rangle \end{pmatrix} \tag{8.10}$$

Now as an example, suppose that

$$|\phi_1\rangle = |\psi_2\rangle \quad |\phi_2\rangle = |\psi_1\rangle \tag{8.11}$$

In this case U is the flip operator: it flips the state $|\psi_1\rangle$ into $|\psi_2\rangle$ and vice versa. Then in the matrix form this is

$$[U] = \begin{pmatrix} 0 & 1 \\ 1 & 0 \end{pmatrix} \tag{8.12}$$

I will drop the square brackets for the matrix notation in the rest of the book. Indeed it is easy to check the flipping operation:

$$\begin{pmatrix} 0 & 1 \\ 1 & 0 \end{pmatrix} \begin{pmatrix} 0 \\ 1 \end{pmatrix} = \begin{pmatrix} 1 \\ 0 \end{pmatrix} \tag{8.13}$$

$$\begin{pmatrix} 0 & 1 \\ 1 & 0 \end{pmatrix} \begin{pmatrix} 1 \\ 0 \end{pmatrix} = \begin{pmatrix} 0 \\ 1 \end{pmatrix} \tag{8.14}$$

We will now put all this knowledge into practice by examining closely two level systems.[1] By the way, the matrix we just wrote down may be familiar to you — it is one of the Pauli matrices (used to describe spin half angular momentum and it appears here because a spin half is a two level system). We'll have more to say about this soon.

8.2 Two Level Systems: Rabi Model

We now specialize all our knowledge of perturbation theory and operator algebra to two level systems. Two level systems are extremely important in physics — there have been three Nobel Prizes awarded to people who studied only them and nothing else! One of those Nobel Prizes was given to Rabi for the problem of a spin half system (basically a two level system) interacting with a magnetic field (or electric — it doesn't make any difference in the maths).

I'd like to spend some time explaining the notation for the Hamiltonian. We will be switching back and forth between the operator ("butterfly") notation $|\alpha\rangle\langle\beta|$, and the matrix notation (both introduced earlier). Proficiency in this manipulation is very important and you will have to practice it by solving problems. The butterfly notation is very convenient as the action of operators becomes apparent. For instance, we have that

$$(|a\rangle\langle b|)|b\rangle = |a\rangle\langle b|b\rangle = |a\rangle \tag{8.15}$$

$$(|a\rangle\langle b|)|a\rangle = |a\rangle\langle b|a\rangle = 0 \tag{8.16}$$

since $\langle b|b\rangle = 1$ and $\langle b|a\rangle = 0$. The operator $|a\rangle\langle b|$ is not Hermitian. Its Hermitian conjugate is given by $(|a\rangle\langle b|)^{\dagger} = |b\rangle\langle a|$. The sum of an operator and its Hermitian conjugate is always itself a Hermitian operator (prove it). The operator which swaps $|a\rangle$ and $|b\rangle$ round (like equation (8.10)) is

$$|a\rangle\langle b| + |b\rangle\langle a| \tag{8.17}$$

---

[1] These are also called *qubits* in literature, which stands for quantum bits. This is in the analogy with the classical bit which is a two state system.

What are the eigenvalues and eigenvectors of this operator? Verify that the eigenvalues are ± 1 and that the (normalized) eigenvectors are

$$|\pm\rangle = \frac{1}{\sqrt{2}}(|a\rangle \pm |b\rangle) \tag{8.18}$$

The purpose of the following is to get used to solving the Schrödinger equation involving a two level atom. Two level systems are ubiquitous in Nature and any physical system can be considered as a collection of two level systems. I will now show you briefly how to solve the evolution of this system and then perform a more detailed calculation to cover some realistic cases.

We will now consider Rabi's problem, but in the optical setting. We have a two level atom interacting with an electromagnetic wave. Consider a two level atom with the corresponding (unperturbed) Hamiltonian

$$\hat{H}_o|i\rangle = \hbar\omega_i|i\rangle, \qquad i = a, b \tag{8.19}$$

interacting with a one mode field of frequency ω linearly polarized, $E(t) = E\cos(\omega t)\hat{\imath}$. Then the time dependent Schrödinger equation

$$i\hbar\frac{\partial\psi}{\partial t}(\mathbf{r}_o, t) = [\hat{H}_o - e\mathbf{r} \cdot \mathbf{E}(\mathbf{r}_o, t)]\psi(\mathbf{r}_o, t) \tag{8.20}$$

has the solution of the form

$$|\psi(t)\rangle = C_a(t)|a\rangle + C_b(t)|b\rangle \tag{8.21}$$

Note that we have the usual dipole field interaction. Here we made what is known as the *dipole approximation*: we assume that the electric field does not vary across the atom. This is true for (visible) light as the wavelength (about $400nm$) is much larger than the atomic size (about $0.1nm$). Using the completeness relation $\sum_{a,b}|i\rangle\langle i| = 1$ the Hamiltonians H_o and H_I can be written as

$$\hat{H}_o = \hbar\omega_a|a\rangle\langle a| + \hbar\omega_b|b\rangle\langle b| \tag{8.22}$$

$$\hat{H}_I = -E(t)(\wp_{ab}|a\rangle\langle b| + \wp_{ba}|b\rangle\langle a|) \tag{8.23}$$

where

$$\wp_{ab} = e\langle a|x|b\rangle = |\wp_{ab}|e^{i\phi} \tag{8.24}$$

is the matrix element of the dipole moment (I am writing things in one dimension now — it is easily generalized to three dimensions).

It has the property $\wp_{ab} = \wp*_{ba}$ and can be written in terms of the amplitude $|\wp_{ab}|$ and phase ϕ. The Rabi frequency is defined as

$$\Omega_R = \frac{|\wp_{ab}|E}{\hbar} \tag{8.25}$$

The equations for the slowly varying amplitudes: $c_i = C_i e^{i\omega_i t}$ in the rotating wave approximation [2] are found substituting equation (8.21) in Schrödinger equation (8.20) using (8.22) and (8.23):

$$\dot{c}_a = i\frac{\Omega_R}{2}e^{-i\phi}c_b e^{i(\omega-\omega_0)t} \tag{8.26}$$

$$\dot{c}_b = i\frac{\Omega_R}{2}e^{i\phi}c_a e^{-i(\omega-\omega_0)t} \tag{8.27}$$

where $(\omega_0 = \omega_a - \omega_b)$. The solution for these equations, with the initial conditions $C_a(0) = 1$ and $C_b(0) = 0$, is

$$c_a(t) = \left[\cos\left(\frac{\Omega}{2}\right) - i\frac{\triangle}{\Omega}\sin\left(\frac{\Omega}{2}\right)\right]e^{\frac{i\triangle t}{2}} \tag{8.28}$$

$$c_b(t) = i\frac{\Omega_R}{\Omega}e^{i\phi}\sin\left(\frac{\Omega t}{2}\right)e^{\frac{-i\triangle t}{2}} \tag{8.29}$$

with

$$\Omega = \sqrt{\Omega_R^2 + (\omega-\omega_0)^2)} \tag{8.30}$$

and $\triangle = \omega - \omega_0$ is the detuning. Check this result by substituting it back into the original equations. (Alternatively, you can differentiate the first of the two equation and substitute in the second one. You can then assume the most general form for the solution of this resulting second order differential equation and obtain the constants of integration). The so-called population inversion, $W(t)$, is an important quantity (as it tells us about the oscillations of the population between the two levels) and is given by

$$W(t) = |c_a(t)|^2 - |c_b(t)|^2 = -\left(\frac{\triangle^2 - \Omega_R^2}{\Omega^2}\right)\sin^2\left(\frac{\Omega t}{2}\right) - \cos^2\left(\frac{\Omega t}{2}\right) \tag{8.31}$$

When the incident radiation is resonant with the transition frequency $\triangle = 0$, the inversion is

Rabi Flopping

$$W(t) = \cos(\Omega t) \tag{8.32}$$

---

[2] In the rotating wave approximation the counter-rotating terms proportional to $e^{\pm i(\omega+\nu)t}$ are ignored because they oscillate much faster than the terms proportional to $e^{\pm i(\omega-\nu)t}$, and therefore "average" to zero much faster.

The population oscillates with Rabi frequency between levels a and b in presence of a resonant field (see Figure 8.1). This behavior is known as the Rabi oscillations or flopping. Note that $E = 0$ implies that $\Omega_R = 0$ and therefore there is no flopping: in the absence of an electromagnetic field the energy levels are stable, so spontaneous emission cannot be explained. The finite life-time of an atom in an excited energy state, which corresponds to a finite line width of the emitted radiation, is only explained by quantum electrodynamic theory in which the interaction of the atom with the (quantized) vacuum field is taken into account. Thus, the semi-classical model predicts that, in the absence of the field, the atom in the excited state does not make transitions to the ground state. Spontaneous emission can be added to the theory phenomenologically introducing a fluctuating electromagnetic background — but this is beyond the scope of the present treatment.

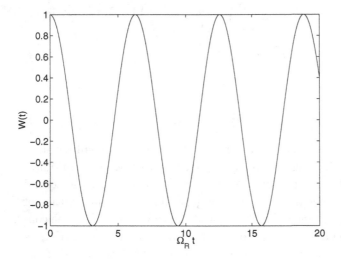

Fig. 8.1   Rabi oscillations of population inversion (defined as the difference in population between the upper and the lower level). It is clear that the population starts in the upper state and then coherently (i.e. through stimulated emission) transfers to the lower level. The process then repeats itself indefinitely.

Here we see the basic quantum mechanical description of emission and absorption of light. The atom, starting from the lower of the two levels, absorbs light and transfers to the upper level (stimulated absorption) and then emits a photon and goes back to the lower level (stimulated emission) and then keeps doing this periodically. The two processes are really mirror images of each other. Emission of light "run backwards" is in fact absorption of light. They are time reversals of each other, which is why we have Einstein's relation between the B coefficients $B_{12} = B_{21}$. And as I said there is no place for spontaneous emission in this treatment.

I am now going to summarize Rabi's treatment and solution and include Fermi's golden rule to obtain the incoherent evolution. So far everything was reversible in this subsection, but we'd also like to include irreversible processes in a more fundamental way. We want to solve the Schrödinger equation with a small dependent perturbation so that the total Hamiltonian is

$$\hat{H}(t) = \hat{H}_0 + \hat{H}'(t) \tag{8.33}$$

The state of a two level system will in general be

$$\Psi(r, t) = c_1(t)\Psi_1(r, t) + c_1(t)\Psi_1(r, t) \tag{8.34}$$

$$= c_1(t)\psi_1(r)e^{-\frac{iE_1}{\hbar}} + c_2(t)\Psi_2(r)e^{-\frac{iE_2}{\hbar}} \tag{8.35}$$

Substituting this into the time dependent Schrödinger equation

$$i\hbar\frac{\partial\Psi}{\partial t} = (\hat{H}_0 + \hat{H}'(t))\Psi \tag{8.36}$$

we obtain

$$i\hbar(\dot{c}_1 + \dot{c}_2) = c_1\hat{H}'\Psi_1 + c_2\hat{H}'\Psi_2 \tag{8.37}$$

This leads to a set of two equations:

$$i\hbar\dot{c}_1 = c_1 H'_{11} + c_2 e^{-i\omega_0 t} H'_{12} \tag{8.38}$$

$$i\hbar\dot{c}_2 = c_1 e^{i\omega_0 t} H'_{21} + c_2 H'_{22} \tag{8.39}$$

where

$$H'_{ij} = \int \psi_i^* \hat{H}' \psi_j dV = \langle\psi_i|H'|\psi_j\rangle \tag{8.40}$$

and

$$\omega_0 = \frac{E_2 - E_1}{\hbar} \tag{8.41}$$

Let's now imagine an electric field interacting with this two level atom

$$\mathbf{E} = E_0\hat{\mathbf{x}}\cos(kz - \omega t) \tag{8.42}$$

But for typical visible light $kz \sim 0.0006$, which is much less than 1 and can be ignored. We assume the following (known as the dipole approximation):

- Ignore B-field effects;   **Dipole**
- Ignore variation of E-field across the atom.   **Approximation**

So we arrive at the following interaction Hamiltonian:

$$\hat{H}' = eE_0\hat{x}\cos(\omega t) \qquad (8.43)$$

Therefore,

$$H'_{12} = \langle 1|\hat{H}'|2\rangle = eE_0 \int \psi_1^*\hat{x}\psi_2 dV \cos\omega t := \hbar\Omega_R \cos\omega t = (H'_{12})^* \qquad (8.44)$$

On the other hand, $H'_{11} = H'_{22} = 0$ as the function x has odd parity. The equations of motion now look easier:

$$i\dot{c_1} = c_2 e^{-i\omega_0 t} H'_{12} \qquad (8.45)$$

$$i\dot{c_2} = c_1 e^{i\omega_0 t} H'_{21} \qquad (8.46)$$

Using the well-known relationship

$$\cos\omega t = \frac{1}{2}(e^{i\omega t} + e^{-i\omega t}) \qquad (8.47)$$

we define

$$D(t) = e^{i(\omega-\omega_0)t} + e^{-i(\omega+\omega_0)t} \qquad (8.48)$$

so that

$$i\dot{c_1} = \frac{1}{2}\Omega_R D(t)c_2 \qquad (8.49)$$

$$i\dot{c_2} = \frac{1}{2}\Omega_R^* D^*(t)c_1 \qquad (8.50)$$

There is still no closed form solution for these coupled equations! We have to approximate. First assume that $c_1(0) = 1$. Then

$$c_1(t) - 1 = \frac{\Omega_R}{2i} \int D(t)c_2(t)dt \qquad (8.51)$$

$$c_2(0) = \frac{\Omega_R^*}{2i} \int D^*(t)c_1(t)dt \qquad (8.52)$$

The first approximation will be that $\omega \approx \omega_0$. This will allow us   **Rotating**
to ignore the rapidly oscillating term $e^{i(\omega_0+\omega)}$ (which averages out   **Wave**
to zero). This approximation, called the Rotating Wave Approxi-   **Approximation**
mation (RWA), is true if

- the time is much higher than the inverse frequency, i.e. $t \gg \omega^{-1}$;
- $\Delta = \omega - \omega_0 \ll \omega_0$, so that the exponential containing $(\omega_0 - \omega)t$ is slowly varying.

Let's look at the special case when $\omega = \omega_0$ (on resonance). Then $\Delta = 0$ and $D(t) = 1$, so that

$$c_1 = \cos \frac{1}{2} \Omega_R t \qquad (8.53)$$

$$c_2 = -i \sin \frac{1}{2} \Omega_R t \qquad (8.54)$$

This means that

$$P_2(t) = |c_2(t)|^2 = \sin^2 \frac{1}{2} \Omega_R t \qquad (8.55)$$

The population is taken from the ground state to the excited state exactly when $\Omega_R t = \pi$, which is called the "pi-pulse" for obvious reasons.

In reality, Rabi oscillations will never be as "clean" as predicted by this formula. The oscillations are will not persist forever, for the same reason that a classical oscillating dipole dies away in the presence of decay. The above formula will look something like (we say something like as the exact form will depend on the exact form of damping and there are many possibilities in reality)

$$P_2(t) = e^{-\gamma t} \sin^2 \Omega_R t \qquad (8.56)$$

so that as $t \to \infty$, the probability to be in the excited state decays to zero.

But as we have already seen we don't need to introduce this decay *ad hoc*. How about using Fermi's golden rule? Let's go back to the equation of evolution of the upper state

$$i\hbar \dot{c}_2 = dE_0 e^{i(\omega - \omega_0)t} c_1 \qquad (8.57)$$

where $\omega_0 = \omega_{12} = \omega_2 - \omega_1$. Assume that, $c_1 \approx 1$, the population in the ground state changes very slowly, i.e. that the probability of a transition is small. Then we can just integrate this equation to obtain

$$c_2(T) - c_2(0) = -i\frac{dE_0}{\hbar} \frac{1}{i(\omega - \omega_0)} [e^{i(\omega - \omega_0)T} - 1] \qquad (8.58)$$

$$= \frac{dE_0}{\hbar} \frac{e^{i(\omega - \omega_0)T} - 1}{\omega - \omega_0} \qquad (8.59)$$

To obtain the probability we need to mod square c_2 (bear in mind that $c_2(0) = 0$):

$$|c_2(t)|^2 = \left[\frac{dE_0 T}{\hbar}\right]^2 \frac{\sin^2 \frac{(\omega - \omega_0)T}{2}}{\left[\frac{(\omega - \omega_0)T}{2}\right]^2} \qquad (8.60)$$

(this is dependent on T^2). The sinc function in this expression peaks at $\omega = \omega_0$, i.e. on resonance. But in reality there will be many frequencies present and the total probability must then be averaged over all frequencies:

$$P = \int \rho(\omega)|c_2(t)|^2 d\omega \approx \rho(\omega_0) \int |c_2(t)|^2 d\omega = \frac{2\pi\rho(\omega_0)d^2 E_0^2}{\hbar^2}T$$
$$(8.61)$$

which is proportional to T, and depends on the density of radiation, dipole strength and the intensity of the field, as usual.

If you are interested in more details I recommend the books by Feynman [Feynman, Leighton and Sands (1989)] and Sakurai [Sakurai (1994)]. These books are full of examples of two level systems embedded in many different settings.

8.3   Other Issues with Two Level Systems

In this section I will also drop "hats" to denote operators. There are many physical examples of (effective) two level systems interacting with some form of a field: a two level atom in a laser field, a two level atom in a magnetic trap, the spin of nucleus in a magnetic field, the polarization state of light, a superposition of a vacuum state and $n = 1$ number state, $|0\rangle + |1\rangle$, interacting with another radiation field and so on. In order to describe any of these situations we need the appropriate interaction Hamiltonian. The most general Hamiltonian for any of these systems can be written as a weighted sum of identity and the three Pauli matrices (together with the identity matrix)

$$H = aI + a_x\sigma_x + a_y\sigma_y + a_z\sigma_z \qquad (8.62)$$

General Two Level Hamiltonian

where the coefficients are real numbers. As all the matrices are Hermitian operators, the resulting operator is Hermitian (and it should be as it's an observable). The coefficients can in general be time dependent. The time evolution of the system is obtained by solving the (potentially time dependent) Schrödinger equation. As an example we can imagine a nuclear spin with two states defined as

$$|0\rangle = \begin{pmatrix} 0 \\ 1 \end{pmatrix} \tag{8.63}$$

$$|1\rangle = \begin{pmatrix} 1 \\ 0 \end{pmatrix} \tag{8.64}$$

Here we are using the matrix representation of operators. Let's see how to go from the butterfly notation to the matrix notation and back. How do we express the Pauli matrices in the operator form? I will first look at σ_x — it can be written as (I know this with hindsight, I've been a quantum physicist for almost 10 years!)

$$\sigma = |0\rangle\langle 1| + |1\rangle\langle 0| \tag{8.65}$$

What are the matrix elements for this operator? It's a two by two matrix (since we have a two level system) and it therefore has four entries: $\langle 0|\sigma_x|0\rangle$, $\langle 0|\sigma_x|1\rangle$, $\langle 1|\sigma_x|0\rangle$ and $\langle 1|\sigma_x|1\rangle$. But

$$\langle 0|\sigma_x|0\rangle = \langle 1|\sigma_x|1\rangle = 0 \tag{8.66}$$
$$\langle 0|\sigma_x|1\rangle = \langle 1|\sigma_x|0\rangle = 1 \tag{8.67}$$

Therefore the matrix is

$$\sigma_x = \begin{pmatrix} \langle 0|\sigma_x|0\rangle & \langle 0|\sigma_x|1\rangle \\ \langle 1|\sigma_x|0\rangle & \langle 1|\sigma_x|1\rangle \end{pmatrix} \tag{8.68}$$

which is, indeed, the familiar Pauli matrix

$$\sigma_x = \begin{pmatrix} 0 & 1 \\ 1 & 0 \end{pmatrix} \tag{8.69}$$

Write other Pauli matrices in the operator form. In the matrix form they are defined as

$$\sigma_y = \begin{pmatrix} 0 & -i \\ i & 0 \end{pmatrix} \quad \sigma_z = \begin{pmatrix} 1 & 0 \\ 0 & -1 \end{pmatrix} \tag{8.70}$$

Imagine now that the magnetic field points initially in the z direction so that

$$H = \mu B \sigma_z \tag{8.71}$$

Suppose that the initial state of the system is

$$|\psi(0)\rangle = |0\rangle + |1\rangle \tag{8.72}$$

Then, by solving Schrödinger equation we obtain (prove it)

$$|\psi(t)\rangle = e^{i\mu B/\hbar}|0\rangle + e^{-i\mu B/\hbar}|1\rangle \tag{8.73}$$

So the spin in some sense processes about the field. This can be seen by computing the average values of σ_x or σ_y (but not σ_z — ask yourself why). Then (prove it)

$$\langle \sigma_x \rangle = \cos(2\mu B/\hbar) \tag{8.74}$$

Imagine now that the field is time dependent. Imagine that we change it from the z direction to the x direction and then to the y direction. This leads to some very interesting (and somewhat surprising) results. In order to see this, let's first talk about the *adiabatic evolution*.

8.4   The Berry Phase

Quantum states are represented as vectors in a complex vector space (with a few other properties we need not worry about here). However, these vectors are only defined up to a unit modulus complex number, or a phase in other words. So, the two states $|\xi\rangle$ and $e^{i\mu}|\xi\rangle$ are indistinguishable as far as quantum mechanics is concerned — no set of measurements can discriminate them. It is interesting that although this statement looks fairly innocent at first sight, it can lead to some very profound conclusions. The route we will take here and that turns out to be very fruitful is to think of this extra phase in a geometric way as well — where we represent the phase as a unit length vector in the complex plane diagram.

Although we think of the amplitudes between two quantum states as fundamental entities in quantum theory, it is only the corresponding probabilities that we can ever observe experimentally. Therefore, given two states $|\psi_i\rangle$ and $|\psi_f\rangle$, we can only measure the mod square of their overlap, $|\langle \psi_i | \psi_f \rangle|^2$. What happens then if we wish to know the relative phase between these two states? Can we define their relative phase in spite of the fact that the absolute phase has no observable consequences? One idea to do so may be to look at the amplitude between the two states in the polar decomposition:

$$\langle \psi_i | \psi_f \rangle = r e^{i\theta_{if}} \tag{8.75}$$

and define the phase θ_{if} as the relative phase between the two states. The main problem with this definition is that the states $e^{i\alpha}|\psi_i\rangle$ and $e^{i\beta}|\psi_f\rangle$, which differ from the original states by an overall arbitrary phase, will have a different relative phases by the amount of $\Delta\theta = \alpha - \beta$. This is not very satisfactory as there are infinitely many choices for $\Delta\theta$ and they all look equally appropriate. We formally say that this definition of phase is *gauge dependent* (meaning "phase dependent").

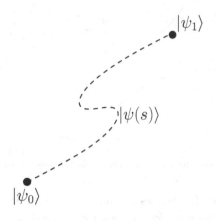

Fig. 8.2 How do we compare phases of two different states when the probability between the states is the only experimentally well-defined quantity in quantum mechanics?

How about defining a path connecting the two states, $|\psi(s)\rangle$, such that when $s = 0$ we have $|\psi_i\rangle$ and when $s = 1$ we have $|\psi_f\rangle$ (see Figure 8.2)? Then we can transport the states $|\psi(s)\rangle$ from the position i to the position f and see how different the final phase is to that of $|\psi_f\rangle$ through *interference*. And if the states $|\psi_f\rangle$ and $|\psi_i\rangle$ transported to $|\psi_f\rangle$ interfere constructively then they are in phase, and if they interfere destructively they are out of phase; the degree of interference can therefore define the phase difference (I will be as vague as possible about what kind of interference this is — I will be much more precise later on). But which path do we choose and how do we know that the transport itself doesn't introduce any additional "twists and turns" in the phases so that we are actually comparing some different phases to the original ones? This, in fact, is a well-known problem in (differential) geometry (see [Misner, Thorne and Wheeler (1973)] for an excellent introduction). Let's state the problem in an abstract setting first (those of you familiar with general relativity will know immediately what I am taking about) and then specialize to quantum geometric phases.

8.4.1 *Parallel transport*

Suppose we have a curved manifold (Figure 8.3) and we have a vector at a point A and another at a point B. What is the angle

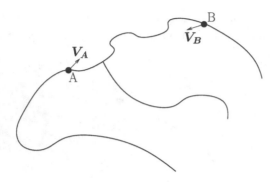

Fig. 8.3 Comparing direction of two vectors at two different points is a non-trivial problem on a curved manifold. We have to somehow transport the vectors to the same point and then see what the angle between them is.

between the two vectors (as phases in quantum mechanics can be thought of as vectors, this question is the same as our original question of relative phase between two quantum states). How do we measure the angle when the two vectors are at different positions? Well, we can transport one of them to the other one and then when they are next to each other the angle is easily measured. But, again, we don't want the transport itself to introduce any additional angles — we'd like it to be as "straight" as possible. The straightest possible path is known as a geodesic and the corresponding evolution along this path is known as the *parallel transport*. To define a parallel transport, let's look at the infinitesimal evolution, from $|\psi(s)\rangle$ to $|\psi(s+ds)\rangle$. If we don't want there to be any twists and turns in the phase, even infinitesimally, then the two states should be in phase. So, we require that

$$\text{Arg}\{\langle\psi(s)|\psi(s+ds)\rangle\} = 0 \qquad (8.76)$$

This is the same as asking that $\langle\psi(s)|\psi(s+ds)\rangle$ be purely real, i.e.

$$\text{Im}\{\langle\psi(s)|\psi(s+ds)\rangle\} = \text{Im}\{\langle\psi(s)|d|\psi(s)\rangle\} = 0 \qquad (8.77)$$

up to second order. But $\langle\psi(s)|d|\psi(s)\rangle$ is purely imaginary anyway (prove it by differentiating $\langle\psi(s)|\psi(s)\rangle = 1$ by ds), and hence the parallel transport condition becomes

$$\langle\psi(s)|d|\psi(s)\rangle = 0 \qquad (8.78)$$

Parallel Transport Rule

If the evolution satisfies this equation, then the phase is parallely transported, which is what we said was required in order to be able to define a relative phase between two states. This definition

of parallel transport is *not* automatically gauge invariant. By this
we mean that if instead of the state $|\psi(s)\rangle$, we use the state

$$|\tilde{\psi}(s)\rangle = e^{i\alpha(s)}|\psi(s)\rangle \tag{8.79}$$

then the parallel transport condition changes by the amount

$$\langle\tilde{\psi}(s)|d|\tilde{\psi}(s)\rangle = \langle\psi(s)|d|\psi(s)\rangle + i\frac{d\alpha}{ds}ds \tag{8.80}$$

as can easily be checked. In order to obtain something that is gauge
invariant we can integrate the expression $\langle\psi|d|\psi\rangle$ over a closed loop,
giving us the expression for the geometric phase, and then exponen-
tiate the result (the integral of $\frac{d\alpha}{ds}$ over a closed loop is 2π, whose
exponential is equal to unity). So, the geometric phase resulting
from the parallel transport is

$$\gamma = \int_i^f \langle\psi(s)|\frac{d}{ds}|\psi(s)\rangle ds \tag{8.81}$$

and its exponential (over a closed loop) is gauge independent, but
not path independent. Incidentally, note that we are requiring
that the phase difference between *every* two infinitesimal points is
zero. So, how come if every infinitesimal difference is zero, on a
finite stretch a non-zero phase builds up? The answer is that the
underlying space is curved and it is the curvature that is reflected in
the phase difference; in fact, the curvature is the phase difference up
to a constant factor. When a quantity vanishes infinitesimally, but
its integral over a finite region doesn't, then this quantity is called
non-integrable. Therefore, geometric phases are a manifestation of
non-integrable phase factors in quantum mechanics. Let's look at
two level systems (spin halfs) to illustrate this point.

8.4.2 *The Bloch sphere*

Two level systems are very conveniently represented on a sphere,
such that all pure states lie on the surface of this sphere and all the
mixed states are inside, as described in section 9.8.1. There is a
one-to-one correspondence between states and points in or on the
Bloch sphere (Figure 8.4). This is because we need at most three
parameters to specify every two level state, as can be seen from the
general density matrix for the systems, that is,

$$\rho = \frac{1}{4}\left(I + \sum_i s_i\sigma_i\right) \tag{8.82}$$

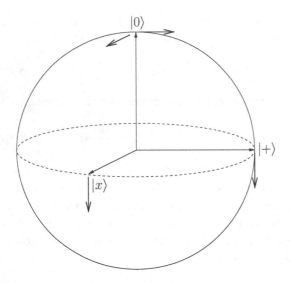

Fig. 8.4 Parallel transport of the phase on the Bloch sphere. The presented evolution can be implemented in many different ways, one of which is the adiabatic Schrödinger equation, as explained in the next section.

where σ_i are the Pauli spin matrices and $s_i = \mathrm{tr}\,\sigma_i\rho$ are the xyz co-ordinates on the sphere. If a state is pure than $s_x^2 + s_y^2 + s_z^2 = 1$, which, as we said, is a point on the surface of the sphere.

Suppose that we now evolve from the state $|0\rangle$ to the state $|+\rangle = |0\rangle + |1\rangle$, then to $|x\rangle = |0\rangle + i|1\rangle$ and finally back to the state $|0\rangle$. On the Bloch sphere we are going from the north pole to the equator, then we move on the equator by an angle of $\pi/2$ and finally we move back to the north pole. What is the corresponding geometric phase? To see this we start with a tangential vector initially at the north pole pointing in some direction (there are infinitely many directions corresponding to infinitely many arbitrary phases to start with). If we now parallely transport this vector along the described path, then we end up with a vector pointing in a different direction to the original one (even though, infinitesimally, the phase vector has always stayed parallel to itself). The angle between the two is $\pi/2$, which is exactly equal to the area covered by the state vector during the transport, or the corresponding solid angle of the transport (the two are just different ways of phrasing the same thing via Stoke's theorem). Therefore, the two orthogonal states $|0\rangle$ and $|1\rangle$ evolve in the following way:

$$|0\rangle \rightarrow e^{i\Omega/2}|0\rangle \qquad\qquad (8.83)$$

$$|1\rangle \rightarrow e^{-i\Omega/2}|1\rangle \qquad\qquad (8.84)$$

where Ω is the solid angle (the factor of half is there because the orthogonal states are π away from each other in the Bloch representation). This is important as it shows that orthogonal states pick up opposite phases of equal magnitude. Finally, we note that the phase can be written as

$$\arg\{\langle 0|+\rangle\langle+|x\rangle\langle x|0\rangle\} \qquad (8.85)$$

This beautiful formula will be the basis of our general formulation of the geometric phase for pure states. From it we find that the space of quantum two level systems is curved. Are there any quantum states whose space is not curved (i.e. flat) and how do we determine the curvature? If you want to measure a curvature at a point, then make a small loop around that point and compare the initial phase of your state and the phase after completing this loop. Suppose that the evolution is

$$|\psi(s_i)\rangle \rightarrow |\psi(s_i + \delta s)\rangle \rightarrow |\psi(s_i + \delta s + \Delta s)\rangle \rightarrow |\psi(s_i)\rangle \qquad (8.86)$$

Then, the final phase difference is

$$
\begin{aligned}
\delta\theta &= \arg\{\langle\psi(s_i)|\psi(s_i+\delta s)\rangle\langle\psi(s_i+\delta s)|\psi(s_i+\delta s+\delta s')\rangle \\
&\quad \times \langle\psi(s_i+\delta s+\delta s')|\psi(s_i)\rangle\} \\
&= \arg\left\{1 + i\left(\frac{\delta}{\delta s}\langle\psi(s_i)|\frac{\delta}{\delta s'}|\psi(s_i)\rangle - \frac{\delta}{\delta s'}\langle\psi(s_i)|\frac{\delta}{\delta s}|\psi(s_i)\rangle\right)\right\} \\
&= \frac{\delta}{\delta s}\langle\psi(s_i)|\frac{\delta}{\delta s'}|\psi(s_i)\rangle - \frac{\delta}{\delta s'}\langle\psi(s_i)|\frac{\delta}{\delta s}|\psi(s_i)\rangle \qquad (8.87)
\end{aligned}
$$

The definition of curvature, K, is $\delta\theta = K\delta A$, where A is the area enclosed. You can convince yourselves that for spin half particles $\delta\theta = \delta A$, which means that the curvature is 1 (it is really $1/r^2$ as for any sphere, but the Bloch sphere has a unit radius). For optical coherent states, on the other hand, $\delta\theta = 0$, indicating that they live on a flat surface (which you can prove as an exercise).

8.4.3 *Implementation*

There is another instance when the time dependent evolution can be handled easily in an analytic way. Suppose that the system starts from an eigenstate of the Hamiltonian. Suppose also that the Hamiltonian changes very slowly (the slowness is measured with respect to the natural frequency of oscillations of the system itself). Then it is natural to expect that the system will stay in the eigenstate of whatever the instantaneous Hamiltonian is. This intuition turns out to be correct and its formal statement is known as the adiabatic theorem. Now we solve a particular instance of adiabatic evolution, which will highlight an issue regarding the quantum phase.

Berry phase is a very interesting phenomenon showing that there is more to the quantum phase than meets the eye. The global phase is, of course, undetectable, but if different quantum states gain different phases then these can be detected through interference. There are two different instances of this relative phase: dynamical and geometric. The dynamical phase, δ, is the usual

$$e^{i\delta} = e^{i\omega t} \tag{8.88}$$

factor which depends on the energy of the system (the dynamical phase should generally be $\exp(\int E(t)dt)$). A particular instance of this geometric phase is Berry's phase, which occurs when the adiabatic theorem is satisfied. I will introduce this through our favorite example, the Mach–Zehnder interferometer. Just to show you how much you have learned so far,[3] I will use neutrons instead of photons and the magnetic (instead of electric) field. The mathematics is exactly the same nevertheless and these experiments have been performed a number of times in reality. The problem we are facing is the following: a neutron beam that is polarized parallel to a uniform magnetic field $\mathbf{B}$ is split into two halves (using a neutron beam splitter). One beam continues through the uniform B-field. The other one passes through a field with the fixed same magnitude, but which changes gradually (i.e. adiabatically)[4] in direction. The two beams are at the end recombined (at the second beam splitter) and the resulting intensity is measured. Our task is to calculate the relative phase change between the two neutron beams at the end of the process.

Suppose that the time dependent magnetic field affecting one half of the beam is

$$\mathbf{B} = B_0(\sin\theta\cos\phi(t)\hat{\mathbf{x}} + \sin\theta\sin\phi(t)\hat{\mathbf{y}} + \cos\theta\hat{\mathbf{z}}) \tag{8.89}$$

where $\phi(t)$ varies adiabatically from 0 to 2π. The neutrons in the other beam experience a constant magnetic field:

$$\mathbf{B} = B_0(\sin\theta\hat{\mathbf{x}} + \cos\theta\hat{\mathbf{z}}) \tag{8.90}$$

---

[3]and to show you that you can solve the Princeton graduate exam problems [Newbury (1991)]!

[4]This should not be confused with the thermodynamic notion of adiabaticity. In thermodynamics this means without heat exchange. Here in optics, or mechanics more precisely, it will mean "slow". The thermodynamic and mechanical notions are, in fact, related once you know statistical mechanics, but this is well beyond the scope of this book.

We are by now already experienced opticians so let's try to guess the result without solving it. The phase comes from precession of the neutron spin around the field. This size of the phase depends on how quickly the spin precesses, i.e. on the magnitude of the magnetic field. But the magnitude of the field in both neutron beams is the same! We can expect the two to be in phase at the end of the process. Wrong! It turns out that there is an additional phase picked up by the beam evolving adiabatically (this was a surprise to Berry who discovered the effect in 1984 in a somewhat different setting). So, our intuition fails here — we need to calculate.

Suppose that the beam takes time T to pass though the interferometer. We are interested in the limit $T \to \infty$, which represents the slow adiabatic evolution. Thus we will solve the Schrödinger equation approximately ignoring terms of the order $1/T^2$ and higher. This is, of course, the time dependent perturbation theory, but like you've never seen before. The equation we are solving is

$$i\hbar \frac{\partial |\psi(t)\rangle}{\partial t} = \mu \mathbf{B}\sigma |\psi(t)\rangle \tag{8.91}$$

where $\mathbf{B}\sigma = B_x \sigma_x + B_y \sigma_y + B_z \sigma_z$. As always assume that

$$|\psi(t)\rangle = a(t)|+\rangle + b(t)|-\rangle \tag{8.92}$$

The point here us that since the Hamiltonian changes (slowly) in time the states $|\pm\rangle$ are also time dependent and follow the Hamiltonian adiabatically (this statement is known as the adiabatic theorem). Therefore

$$|+\rangle = \cos \frac{\theta}{2}|0\rangle + e^{\phi(t)} \sin \frac{\theta}{2}|1\rangle \tag{8.93}$$

$$|-\rangle = \sin \frac{\theta}{2}|0\rangle + e^{-\phi(t)} \cos \frac{\theta}{2}|1\rangle \tag{8.94}$$

Let's look at

$$\frac{\partial |\psi(t)\rangle}{\partial t} = \frac{da}{dt}|+\rangle + a\frac{\partial |+\rangle}{\partial t} + \frac{db}{dt}|-\rangle + b\frac{\partial |-\rangle}{\partial t} \tag{8.95}$$

If we multiply by $\langle+|$ only the first two terms survive. The third term is orthogonal and the fourth is of the order $1/T^2$ since

$$b\frac{\partial |-\rangle}{\partial t} = b\dot{\phi}\frac{\partial |-\rangle}{\partial \phi} \tag{8.96}$$

and $\dot{\phi} = 2\pi/T$. The Schrödinger equation becomes

$$i\hbar \frac{da}{dt} + a(t)\langle+|i\hbar\frac{\partial}{\partial t}|+\rangle = -E_0 a(t) \tag{8.97}$$

And here is the trick: the only way that we can solve this is to assume that $a(t) = e^{i\gamma(t)/\hbar}$ (this is an extra phase in addition to the "usual" dynamical phase iEt). We obtain

$$\frac{d\gamma}{dt} - E_0 = \langle +|i\hbar\frac{\partial}{\partial t}|+\rangle \tag{8.98}$$

$$= \hbar\sin^2\frac{\theta}{2}\dot{\phi} \tag{8.99}$$

where $E_0 = 2\mu B/\hbar$. We integrate this expression from 0 to T:

$$\frac{1}{\hbar}(\gamma(T) - \gamma(0) - E_0 T) = -2\pi\sin^2\frac{\theta}{2} \tag{8.100}$$

The other beam also gets $E_0 T/\hbar$ phase so that the phase difference between the two beams is

Berry
Phase

$$\Delta = -2\pi\sin^2\frac{\theta}{2} \tag{8.101}$$

The intensity is therefore proportional to

$$I \propto |1 + e^{i\Delta}|^2 \propto \cos^2\frac{\Delta}{2} = \cos^2\left(\pi\sin^2\frac{\theta}{2}\right) \tag{8.102}$$

(this is just interference from two different beams as before). Thus the two beams are not in phase as they emerge. The phase difference between the two beams is of a purely geometric origin, i.e. it only depends on the geometry of the motion executed by the field in the parameter space.

8.4.4 *Generalization of the phase*

I will now briefly generalize the concept of the geometric phase.[5] The states of a quantum system are usually described as being represented by vectors of norm 1 ($|\langle\psi|\psi\rangle|^2 = 1$) in a complex Hilbert space $\mathcal{H}$. However, there is redundancy in this description since the state $|\psi\rangle$ is physically indistinguishable from the state $e^{i\phi}|\psi\rangle$. It is therefore convenient to consider the projective space $\mathcal{P}$, in which indistinguishable physical states are grouped into equivalence classes under the relation $|\psi\rangle \sim re^{i\phi}|\psi\rangle$ for any $r > 0$ and real ϕ. The associated projection map is

$$\begin{aligned}\Pi : \mathcal{H} &\to \mathcal{P}\\ |\psi\rangle &\mapsto [|\psi\rangle] = \{|\psi'\rangle : |\psi'\rangle = re^{i\phi}|\psi\rangle\}\end{aligned} \tag{8.103}$$

---

[5]If you are interested in this subject you can read my review *Geometric Phases and Topological Quantum Computation*, Int. J. Quant. Info. **1**, 1 (2003).

If a system undergoes a cyclic evolution, the ket representing the system state traces out a path, $\mathcal{C} : [0, \tau] \longrightarrow \mathcal{H}$, where $\Pi(\mathcal{C})$ is a closed curve in $\mathcal{P}$, as illustrated in Figure 8.5. In other words, the initial and final states should be on the same ray in $\mathcal{H}$, but may be related by a phase, $e^{i\phi}$. We will measure this phase with respect to a reference curve in $\mathcal{H}$: for each point $|\psi(t)\rangle$ on $\mathcal{C}$, we can choose a smoothly varying representative $|\tilde{\psi}(t)\rangle$ from $\Pi(\psi(t))$ in such a way that $|\tilde{\psi}(0)\rangle = |\tilde{\psi}(\tau)\rangle$. We can then write

$$|\psi(t)\rangle = e^{if(t)}|\tilde{\psi}(t)\rangle \qquad (8.104)$$

so that the phase change of $|\psi(0)\rangle$ associated with the cyclic evolution is given by $\phi = f(\tau) - f(0)$.

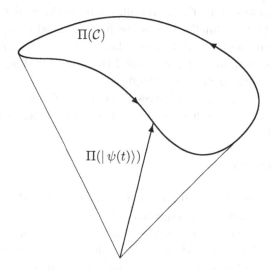

Fig. 8.5 Cyclic State Evolution. The geometric part of the phase generated by this evolution is proportional to the area covered by the state vector.

The time evolution of a quantum system is governed by the Schödinger equation

$$i\hbar \frac{d}{dt} |\psi(t)\rangle = H(t) |\psi(t)\rangle \qquad (8.105)$$

where $H(t)$ is the Hamiltonian. Note that we are not making the adiabatic assumption. Substituting equation (8.104) into the above, rearranging and multiplying by $\langle \psi(t) |$ gives the following:

$$\frac{df(t)}{dt} = -\frac{1}{\hbar} \langle \psi(t) \,|\, H \,|\, \psi(t) \rangle + i \langle \tilde{\psi}(t) | \frac{d}{dt} | \tilde{\psi}(t) \rangle \qquad (8.106)$$

or, when integrated,

$$\phi = -\frac{1}{\hbar} \int_0^\tau \langle \psi(t) \,|\, H \,|\, \psi(t) \rangle \, dt + i \int_0^\tau \langle \tilde{\psi}(t) | \frac{d}{dt} | \tilde{\psi}(t) \rangle dt \qquad (8.107)$$

Thus, ϕ can be decomposed into a dynamical phase

$$\delta = -\frac{1}{\hbar} \int_0^\tau \langle \psi(t) \,|\, H \,|\, \psi(t) \rangle \, dt \qquad (8.108)$$

which depends on the Hamiltonian, and a geometric phase **Geometric Phase**

$$\gamma = i \oint_C \langle \tilde{\psi} | d | \tilde{\psi} \rangle \qquad (8.109)$$

which depends only on the path C; γ is independent of the rate at which $| \psi(t) \rangle$ progresses along C, the Hamiltonian, or the choice of reference $\{ | \tilde{\psi} \rangle \}$. And, so finally, this is the most general formula for the phase in any setting. For those who would like to know more about geometric phases, I recommend the book by Wilczek and Shapere [Wilczek and Shapere (1990)].

At the end I'd like to discuss phases in quantum mechanics a bit more. I want to show you how to derive electromagnetism from quantum mechanics!

8.5   Gauge Principle

I would now like to discuss the interaction between light and matter, but from a very different perspective. We have seen that the interaction between the two happens via the atomic dipole and electric field coupling. But where does this Hamiltonian come from? The deepest and most beautiful answer we have is called the gauge invariance principle. This principle will show us how to derive electromagnetism from quantum mechanics! Recall that the basic equation governing all the (non-relativistic) behavior in Nature is given by the Schrödinger equation:

$$\frac{1}{2m}((-i\nabla)^2 + V)\Psi(x,t) = i\hbar \frac{\partial}{\partial t} \Psi(x,t) \qquad (8.110)$$

We know that the physical meaning of the wave function comes from its mod square being the probability density of finding the system at position x and time t. Therefore if we introduce a phase to the wave function $\Psi(x,t) \to e^{i\theta} \Psi(x,t)$, physics will not change as the mod square stays the same. Also, by plugging the "phase shifted" wave function into the Schrödinger equation we see that it also satisfies it. So, the Schrödinger equation remains invariant under constant phase shifts — this is known as the global gauge invariance. So far so good. Let's look now at the case when the phase shift is space and time dependent. The probability is still unchanged. Fine. But the Schrödinger equation changes as the phase now "sees" the space and time derivatives. But this is not completely satisfactory. The reason is that changing a phase at (x_1, t_1) should not be able to affect anything at (x_2, t_1), for example. Relativistically we should expect to be able to change the two phases at two different points independently since the change at one point can travel to the other point at most as fast as light. Indeed, why should we not be able to affect different space time points differently? Any (instantaneous) communication would be seriously acausal. The only way we can imagine that different points are "connected" would be if there was a force between the two points, "informing" one point about what happens at the other point. If we, therefore, require that the Schrödinger equation is invariant under the "local" phase change, we can only maintain this condition providing that we implement the following transformations:

$$-i\nabla \rightarrow -i\nabla - \nabla\theta \qquad (8.111)$$

$$V \rightarrow V - \frac{\partial\theta}{\partial t} \qquad (8.112)$$

But this looks deceptively like the gauge transformation of the vector potential in electromagnetic theory. There, Maxwell's equations are invariant under the following change of the vector potential

$$A \rightarrow A' = A + \nabla\theta \qquad (8.113)$$

providing that we also change

$$V \rightarrow V' = V - \frac{\partial\theta}{\partial t} \qquad (8.114)$$

Therefore, the appropriate Schrödinger equation should be

$$\frac{1}{2m}((-i\hbar\nabla - eA)^2 + V)\Psi(x, t) = i\hbar\frac{\partial}{\partial t}\Psi(x, t) \qquad (8.115)$$

and is invariant under the local phase change providing that the vector potential A and the scalar potential V change the way that the electromagnetic vector and scalar potential do. Thus, the local gauge invariance of the Schrödinger equation implies the electromagnetic field. The electromagnetic field exists in order to mediate wave function phases between different space time points! Now, it turns out that all fundamental forces of nature can be described this way apart from gravity. Therefore this offers a very unified way of looking at physics.

Let's now look at light matter interaction Hamiltonian to try to justify our previous choice. The Hamiltonian is

$$H = \frac{1}{2m}((-i\hbar\nabla - eA)^2) = -\frac{1}{2m}(\hbar^2\nabla^2 + e(-i\nabla)A - e^2A^2) \quad (8.116)$$

So, up to the first order, the light matter interaction Hamiltonian is proportional to the term $\hat{p}\hat{A}$. Now we show that this is equivalent to the $-\hat{d}\hat{E}$ Hamiltonian which we have been using throughout the book. I will give you a "hand waving" picture that can be made rigorous. The interaction Hamiltonian can be written as

$$V_{int} = -epA = -epA_0 e^{i\omega nx/c} \quad (8.117)$$

But in quantum optics the dipole approximation applies meaning that the wavelength of light is much larger than the typical atomic dimensions (a couple of Ängstroms). So, let's use the Taylor expansion on the above:

$$-epA_0\left(1 + \frac{i\omega nx}{c} + \cdots\right) \quad (8.118)$$

So, taking into account the lowest order only, the transition element between some two levels (the only thing we ever want and need to calculate in quantum mechanics) is

$$\langle f|pA|i\rangle \approx \langle f|p|i\rangle \quad (8.119)$$
$$\propto \langle f|[x, H_0]|i\rangle \quad (8.120)$$
$$\propto \langle f|x|i\rangle \quad (8.121)$$

where $[x, H_0] = xH_0 - H_0x$ is the usual commutator between the position operator and the free Hamiltonian, which is equal to the momentum operator as used in the second line of the above proof. So we may well use the interaction exE instead pA.

Final Thought: So we have now through studies of phases arrived at a deep understanding of the electromagnetic interactions. I'd like to summarize the necessity of the electromagnetic field as a consequence of the gauge invariance in a slightly different way, involving the superposition principle. Suppose that the process

$$|\psi_i\rangle \rightarrow |\psi_f\rangle \quad (8.122)$$

can happen in two different ways, via $|\psi_1\rangle$ and $|\psi_2\rangle$. Suppose further that these two acquire different phases for some reason

$$e^{i\theta(p_1)}|\psi_1\rangle + e^{i\theta(p_2)}|\psi_2\rangle \qquad (8.123)$$

which depend on their paths p_1 and p_2. The relative phase is now

$$e^{i\theta(p_1)}\left(|\psi_1\rangle + e^{i(\theta(p_2)-\theta(p_1))}|\psi_2\rangle\right) \qquad (8.124)$$

If we have that

$$\theta(p_i) = \int_{p_i} \theta(x)\mathbf{dx} \qquad (8.125)$$

so that

$$\theta(p_2) - \theta(p_1) = \oint \theta(x)\mathbf{dx} \qquad (8.126)$$

If we don't want any phase dependence we have to "invent (design) a field" that cancels it — hence the electromagnetic field. The weak and strong force fields can be derived in this way as well, but asking for a more complicated gauge independence (but we cannot obtain gravity within this formalism)!

Chapter 9

Field Quantization

We have been loosely stating at various stages that a quantum field is represented by a collection of simple harmonic oscillators. Where does this come from? Let's take another look at Maxwell's equations — this is the best electromagnetic field description we have at this stage. Can we quantize it? In fact, what does it mean to quantize it? We have six numbers associated with every point in space and time: three represent the value of the electric field (the x, y and z components) and three represent the value of the magnetic field. That's all there is to light (and electromagnetism) classically! But, wait a minute. Can we know all these values simultaneously? Classically — yes, quantum mechanically — no. Let's see why we cannot specify the electric and magnetic fields simultaneously. The precise statement is that there is a limit to accuracy in the two fields in the orthogonal directions and the original reasoning is due to Bohr, I believe (many people have worked on this including Landau, Peirles and Resenfeld, but all of these came from the Copenhagen's school of Bohr's — see [Heisenberg (1930)]). How do we measure the electric field? We put a test charge in the field. The direction in which it moves tells us the direction of the field, while the amount by which it moves tells us its strength. To estimate the magnetic field, on the other hand, we need to measure the velocity of the same test charge. But if the magnetic field is perpendicular to the electric, then the velocity is in the direction of the position measurement we need for the electric field! So, measuring the electric and magnetic fields boils down to measuring the position and momentum of a test charge. But, Heisenberg's uncertainty relation already tells us that this is not possible to do simultaneously! So, this indicates that we cannot specify the electromagnetic field simultaneously. Note that this logic, although quite believable, is by no means a proof of the uncertainty in the electromagnetic field. First of all, maybe there is another way which does not involve a single test charge. Can we use two charges, or more? Secondly, uncertainty cannot really be proved directly — it is inferred from experiments and from observations of its consequences. One of the consequences of non-commutability of the electric and the magnetic field is that there are photons. Another one is that the vacuum has in some sense a non-zero amount of energy. There are other consequences, and I will discuss them in the remaining part.

How do we mathematically describe the fact that the electric and the magnetic field values cannot be specified simultaneously quantum mechanically? We have to elevate the electric and magnetic field from the status of mere vectors, to the status of operators! Here is the standard, textbook, way of doing this [Mandel and Wolf (1995)]. (Beware: the fact that it's straight out of a textbook doesn't mean it's easy!)

As you know the electric and the magnetic fields can be expressed via a vector potential, which also satisfies the wave equation (just like the B and E fields themselves)

$$\nabla^2 \mathbf{A} = \frac{1}{c^2} \frac{\partial^2 \mathbf{A}}{\partial t^2} \tag{9.1}$$

where the magnetic field is the curl of the vector potential, i.e. $\mathbf{B} = \nabla \wedge \mathbf{A}$, and the electric field is the rate of change of the vector potential, i.e. $\mathbf{E} = \partial \mathbf{A}/\partial t$. A formal solution to this is given by a Fourier series

$$\mathbf{A} = \sum_{\mathbf{k}} (\mathbf{A_k}(t)e^{i\mathbf{kr}} + \mathbf{A_k^*}(t)e^{-i\mathbf{kr}}) \tag{9.2}$$

Inserting this back into the original wave equation we see that each Fourier component has to itself satisfy the same wave equation, i.e.

$$\nabla^2 \mathbf{A_k} = \frac{1}{c^2} \frac{\partial^2 \mathbf{A_k}}{\partial t^2} \tag{9.3}$$

We now want to make A look as much as possible like the simple harmonic oscillator. How can we do that? First of all, let's express the E and B fields using the A potential. We have, assuming that $A(t) = Ae^{i\omega t}$,

$$\mathbf{E_k} = i\omega_{\mathbf{k}}(\mathbf{A_k}e^{-i\omega_{\mathbf{k}}t+i\mathbf{kr}} - \mathbf{A_k^*}e^{i\omega_{\mathbf{k}}t-i\mathbf{kr}}) \tag{9.4}$$

$$\mathbf{B_k} = i\mathbf{k} \wedge (\mathbf{A_k}e^{-i\omega_{\mathbf{k}}t+i\mathbf{kr}} - \mathbf{A_k^*}e^{i\omega_{\mathbf{k}}t-i\mathbf{kr}}) \tag{9.5}$$

(The first equation is a time derivative of the vector potential, while the second is its curl.) Now, the energy of a single mode k is given by

$$W_{\mathbf{k}} = \frac{1}{2} \int (\epsilon_0 \mathbf{E}_{\mathbf{k}}^2 + \mu_0^{-1} \mathbf{B}_{\mathbf{k}}^2) dV \qquad (9.6)$$

which is in terms of A_ks

$$W_{\mathbf{k}} = 2\epsilon_0 V \omega_{\mathbf{k}}^2 \mathbf{A}_{\mathbf{k}} \mathbf{A}_{\mathbf{k}}^* \qquad (9.7)$$

But this looks very much like the energy of a quantum harmonic oscillator ($\hbar \omega a^\dagger a$)! To see the exact correspondence, we define

$$\mathbf{A}_{\mathbf{k}} = (4\epsilon_0 V \omega_{\mathbf{k}}^2)^{-1/2} (\omega_{\mathbf{k}} \mathbf{Q}_{\mathbf{k}} + i \mathbf{P}_{\mathbf{k}}) \epsilon_{\mathbf{k}} \qquad (9.8)$$
$$\mathbf{A}^*{}_{\mathbf{k}} = (4\epsilon_0 V \omega_{\mathbf{k}}^2)^{-1/2} (\omega_{\mathbf{k}} \mathbf{Q}_{\mathbf{k}} - i \mathbf{P}_{\mathbf{k}}) \epsilon_{\mathbf{k}} \qquad (9.9)$$

where $\epsilon_{\mathbf{k}}$ is the polarization of light. So, finally

$$W_{\mathbf{k}} = \frac{1}{2} (P_{\mathbf{k}}^2 + \omega_{\mathbf{k}}^2 Q_{\mathbf{k}}^2) \qquad (9.10)$$

which is a unit mass simple harmonic oscillator! Note that all this so far is purely classical. But, we have arrived at something familiar — a harmonic oscillator, and we know how to quantize a harmonic oscillator.

9.1 Quantum Harmonic Oscillator

I would like to remind you of the way of solving the quantum harmonic oscillator. The Hamiltonian is in this case given by

$$\hat{H} = \frac{1}{2} (\hat{p}^2 + \hat{q}^2) \qquad (9.11)$$

For simplicity, I have omitted the frequency ω^2 in front of $\hat{q}^2$ by setting it equal to unity. (I will put it back in when it becomes important!) It is customary to perform a change of variables

Creation and Annihilation Operators

$$\hat{a} = \frac{1}{\sqrt{2}} (\hat{p} + i\hat{q}) \qquad (9.12)$$

$$\hat{a}^\dagger = \frac{1}{\sqrt{2}} (\hat{p} - i\hat{q}) \qquad (9.13)$$

We see that $(p^2 + q^2) = a^\dagger a + a a^\dagger$. From the commutator of p and q, $[q, p] = i\hbar$, we can deduce

$$[\hat{a}, \hat{a}^\dagger] = 1 \qquad (9.14)$$

so that

$$H = \hbar \omega (\hat{a}^\dagger \hat{a} + \frac{1}{2}) \qquad (9.15)$$

I have now put back the $\hbar\omega$ (otherwise dimensions would be wrong). We can now compute two more commutators that will allow us to compute the spectrum of the above light Hamiltonian. We have

$$[H, a] = \hbar\omega[a^\dagger, a]a = -\hbar\omega a \qquad (9.16)$$

and

$$[H, a^\dagger] = \hbar\omega a^\dagger[a, a^\dagger] = \hbar\omega a^\dagger \qquad (9.17)$$

Now, suppose that the eigenvalue equation for the nth level is given by (this is completely general, by definition of the eigenvalue equation)

$$H|\psi_n\rangle = E_n|\psi_n\rangle \qquad (9.18)$$

and we have that

$$[H, a]|\psi_n\rangle = (Ha - aH)|\psi_n\rangle = (H - E_n)a|\psi_n\rangle = -\hbar\omega a|\psi_n\rangle \quad (9.19)$$

Therefore,

$$H(a|\phi_n\rangle) = (E_n - \hbar\omega)|\psi_n\rangle \qquad (9.20)$$

and so the action of the operator a is to lower the energy of the nth state by one unit of $\hbar\omega$. It is thus called the lowering operator. An analogous equation can be derived for $a^\dagger$, whose action is to raise the value of energy by one unit of $\hbar\omega$. It can be derived that

$$H(a^\dagger|\phi_n\rangle) = (E_n + \hbar\omega)|\psi_n\rangle \qquad (9.21)$$

and $a^\dagger$ is therefore known as the raising operator.

We already know that we are discussing is light which consists of photons. In fact, the operator $\hat{a}^\dagger a$ is a Hermitian operator and counts the number of quanta (photons, if you like) present in the oscillator. Mathematically, this is expressed as

$$\hat{a}^\dagger \hat{a}|n\rangle = n|n\rangle \qquad (9.22)$$

where $|n\rangle$ is the nth eigenstate of the Hamiltonian. This physically means that if we were to measure the number of photons in the state $|n\rangle$, we would obtain n photons with unit probability. Therefore the amount of energy in the nth state is

$$H|n\rangle = \hbar\omega\left(\hat{a}^\dagger\hat{a} + \frac{1}{2}\right)|n\rangle = \hbar\omega\left(n + \frac{1}{2}\right)|n\rangle \qquad (9.23)$$

So even when there are no photons, i.e. $n = 0$, there is still some energy left over, known as the zero point energy (see Figure 9.1), of $\hbar\omega/2$.

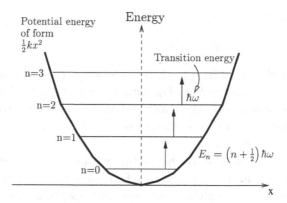

Fig. 9.1 The spectrum of a quantum harmonic oscillator. Here a vibrating molecule is representing; for us, the most important result is that light behaves in exactly the same way as this molecule — it is a quantum harmonic oscillator and has the same type of spectrum as above. Notice that when there are no photons there is still some residual energy left over of $\hbar\omega/2$. This is the so-called zero point energy and is responsible, among other things, for spontaneous emission and the Casimir effect.

Zero point energy will have some astonishing effects, such as the Casimir force, as we shall see later in the book. In general, however, the state of the harmonic oscillator can be any linear superposition of number states, like

$$|\Psi\rangle = \sum_n a_n |n\rangle \tag{9.24}$$

The average energy in this state is

$$\langle\Psi|H|\Psi\rangle = \sum_n a_n^* \langle n|H \sum_m a_m |m\rangle \tag{9.25}$$

$$= \sum_{n,m} a_n^* a_m \left(\hbar\omega\left(m + \frac{1}{2}\right)\right) \langle n|m\rangle \tag{9.26}$$

$$= \frac{1}{2}\hbar\omega \sum_n |a_n|^2 + \hbar\omega \sum_n |a_n|^2 n \tag{9.27}$$

$$= \frac{1}{2}\hbar\omega + \hbar\omega \sum_n |a_n|^2 n \tag{9.28}$$

This result is expected from our previous discussion. The energy of n photons is $1/2 + n\hbar\omega$ and the probability of having n photons is $p_n = |a_n|^2$, hence the average $\hbar\omega \sum_n p_n n$.

But, what is the exact action of a and $a^\dagger$ on the number state? We are now in a position to derive it. Here is how. We know that

$$\langle n|a^\dagger a|n\rangle = n \tag{9.29}$$

But we know that $\langle n|a^\dagger = (a|n\rangle)^\dagger$, so that

$$a|n\rangle = \sqrt{n}|m\rangle \tag{9.30}$$

But $a^\dagger(a|n\rangle) = n|n\rangle$, so that $m = n-1$. Therefore, a is the lowering operator as it reduces the number of quanta in the state, and $a^\dagger$ is the raising operator as it raises the number of quanta by 1 unit. We have actually already seen this as the action of a leads to energy reduction from E_n to $E_n - \hbar\omega$.

What is the correspondence of this picture with the wave function Schrödinger picture that you are very familiar with? Remember that wave functions are always expressed in terms of position co-ordinates. This we obtain by taking the following product,

$$\psi(x) = \langle x|\psi\rangle \tag{9.31}$$

And we know from before that, for example,

$$\langle x|0\rangle \sim e^{x^2/2} \tag{9.32}$$

Higher states can be obtained by applying $a^\dagger$ to the ground state, i.e.

$$|n\rangle = \frac{1}{n!}(a^\dagger)^n|0\rangle \tag{9.33}$$

I'd now like to show you a very beautiful way of using the Dirac bracket notation to capture the operation of the raising and lowering operators. Suppose we look at the construction of the type

$$|n+1\rangle\langle n| \tag{9.34}$$

When this acts on the state $|n\rangle$ it raises it to $|n+1\rangle$, as $\langle n|n\rangle = 1$, like so:

$$|n+1\rangle\langle n|(|n\rangle) = |n+1\rangle\langle n|n\rangle = |n+1\rangle \tag{9.35}$$

However, when this operator acts on any other state other than n we obtain 0. Thus the raising operator can be written as

$$\sqrt{1}|1\rangle\langle 0| + \sqrt{2}|2\rangle\langle 1| + \cdots + \sqrt{n+1}|n+1\rangle\langle n| + \cdots \tag{9.36}$$

It is now straightforward to write the lowering operator:

$$\sqrt{0}|0\rangle\langle 0| + \sqrt{1}|0\rangle\langle 1| + \cdots + \sqrt{n}|n-1\rangle\langle n| + \cdots \tag{9.37}$$

The product of the two can now be seen to be

$$0|0\rangle\langle 0| + 1|1\rangle\langle 1| + \cdots + n|n\rangle\langle n| + \cdots \tag{9.38}$$

which is the number operator (try acting on some state $|m\rangle$ to convince yourself of this).

A third way of writing this is the matrix (Heisenberg) form:

$$a = \begin{pmatrix} 0 & 0 & 0 & 0 & \dots \\ \sqrt{1} & 0 & 0 & 0 & \dots \\ 0 & \sqrt{2} & 0 & 0 & \dots \\ 0 & 0 & \sqrt{3} & 0 & \dots \\ \vdots & \vdots & \vdots & \vdots & \ddots \end{pmatrix}, \quad a^\dagger = \begin{pmatrix} 0 & \sqrt{1} & 0 & 0 & \dots \\ 0 & 0 & \sqrt{2} & 0 & \dots \\ 0 & 0 & 0 & \sqrt{3} & \dots \\ 0 & 0 & 0 & 0 & \dots \\ \vdots & \vdots & \vdots & \vdots & \ddots \end{pmatrix}.$$

The number operator is, therefore,

$$a^\dagger a = \begin{pmatrix} 0 & 0 & 0 & 0 & \dots \\ 0 & 1 & 0 & 0 & \dots \\ 0 & 0 & 2 & 0 & \dots \\ 0 & 0 & 0 & 3 & \dots \\ \vdots & \vdots & \vdots & \vdots & \ddots \end{pmatrix}$$

diagonal in the number state basis with the eigenvalues being the number of quanta in the corresponding number state.

Given this quantum harmonic oscillator representation of a radiation field, what is a photon physically?

9.2 What Are Photons?

Here is a simpler way of understanding the quantization of the electromagnetic field. Let us imagine a one-dimensional cavity with perfectly reflecting mirrors, placed at $z = 0$ and $z = L$, filled with a monochromatic electromagnetic field. In order to satisfy the imposed boundary conditions, i.e. the electric field has to vanish at the mirrors, whereas the magnetic field strength reaches its maximum, we must have

$$E_x(t) = \sqrt{\frac{2\omega^2}{\epsilon_0 L}} q(t) \sin kz \tag{9.39}$$

$$H_y(t) = \frac{\epsilon_0}{k} \sqrt{\frac{2\omega^2}{\epsilon_0 L}} \dot{q}(t) \cos kz \tag{9.40}$$

where $kL = n\pi$, n being an integer. Our choice of writing $E_x(t)$ and $H_y(t)$ in this particular form is immediately justified by observing the overall energy in the cavity, per frequency, which is given by

$$W = \frac{1}{2} \int_0^L (\epsilon_0 E_x^2 + \mu_0 H_y^2) dz \tag{9.41}$$

Substituting the expressions for $E_x(t)$ and $H_y(t)$ and writing $p = \dot{q}$ we obtain (prove it)

$$W = \frac{1}{2}(p^2 + \omega q^2) \tag{9.42}$$

where we have used the fact that $c = 1/\sqrt{\mu_0 \epsilon_0}$ and $\omega = kc$. This is a familiar form of the energy of a unit mass harmonic oscillator at frequency ω. It strongly suggests that the electromagnetic field should be quantized via canonically conjugate variables q and p (a more rigorous analysis shows that this is a correct way of proceeding). It will, therefore, exhibit all the properties of the quantized mechanical harmonic oscillator (QMHO). In particular, the energy will be quantized with the eigenvalues having the form

$$E_n = \hbar\omega \left(n + \frac{1}{2} \right), \quad n = 0, 1, 2, \ldots \tag{9.43}$$

Invoking raising and lowering operators, $\hat{a}^\dagger$ and $\hat{a}$, we can write

$$q \rightarrow \hat{q} = \sqrt{\frac{\hbar}{2\omega}}(\hat{a} + \hat{a}^\dagger) \tag{9.44}$$

$$p \rightarrow \hat{p} = -i\sqrt{\frac{\hbar\omega}{2}}(\hat{a} - \hat{a}^\dagger) \tag{9.45}$$

Therefore, the quantized electric field, in a one-dimensional cavity can be written using $\hat{a}^\dagger$ and $\hat{a}$ as

$$\hat{E}_x(t) = \sqrt{\frac{\hbar\omega}{\epsilon_0 L}}\, \overrightarrow{x}(\hat{a} + \hat{a}^\dagger) \sin kz \tag{9.46}$$

Quantized Electric Field

where $\overrightarrow{x}$ is a unit vector in the x direction, and $\sqrt{\frac{\hbar\omega}{\epsilon_0 L}}$ is the so-called electric field per photon. We note in passing that since $\hat{q}$ and $\hat{p}$ obey the Heisenberg commutation relations, i.e. $[\hat{q}, \hat{p}] = -i$, so will the quantized E and H. Thus, similar uncertainty relations to the ones obeyed by $\hat{q}$ and $\hat{p}$ are also obeyed by the quantized E and H fields. In what follows we deal exclusively with quantum systems, so that we omit hats, which are superfluous.

So what can we say about photons — particles of light — now that we know how quantum mechanics deals with them mathematically? In the strictest sense, a photon is a single excitation of a single mode of the quantized electromagnetic field. Therefore it is not exactly a particle in the conventional sense of the word as it is not necessarily localized. In spite of that, a single quantum of light will register only a single click from a detector which is very much a particle-like behavior. In the ultimate description of

Nature we have — quantum field theory — every particle behaves in this way. There is a field associated with every quantum system and particles are just excitations of vibrations of that field.

9.3   Blackbody Spectrum from Photons

I would like now to show a very simple argument that can be used to derive the blackbody spectrum, but now from this quantized field picture of photons that we have just been analyzing. Suppose that we have n photons present in a cavity. The action of creating another photon gives us the state $a^\dagger|n\rangle$. The probability to have another photon emitted is then described by the mod square of this:

$$\langle n|aa^\dagger|n\rangle = n + 1 \qquad (9.47)$$

The probability to absorb a photon is, on the other hand, given by

$$\langle n|a^\dagger a|n\rangle = n \qquad (9.48)$$

Now, if we are talking about the average number of photons emitted then we will use $\langle n\rangle$ instead of n. The total rate of emission is then

$$N_2(\langle n\rangle + 1) \qquad (9.49)$$

where N_2 is the number of excited atoms, while the total rate of absorption becomes

$$N_1\langle n\rangle \qquad (9.50)$$

where N_1 is the number of atoms in the ground state. Note the difference between emission and absorption of +1. This is the spontaneous emission term, which is absent in the absorption rate. Equating these two rates should give us the equilibrium average photon distribution, which is the same as Planck's. Indeed, from the equation

$$N_2(\langle n\rangle + 1) = N_1\langle n\rangle \qquad (9.51)$$

it follows that

$$\langle n\rangle = \frac{1}{\frac{N_1}{N_2} - 1} = \frac{1}{e^{\hbar\omega/kT} - 1} \qquad (9.52)$$

which is just the blackbody spectrum! And this derivation is probably the most rigorous way of deriving it with the full machinery of field quantization available to us.

Now I'd like to discuss one of the most bizarre effects of the existence of quantum vacuum: Casimir's force.

9.4   Quantum Fluctuations and Zero Point Energy

The state $|0\rangle$, referred to as the vacuum state, is the state of lowest energy and corresponds to the state for which all the occupation numbers $n_{\mathbf{k}}$ are zero. Although no excitations are present, the total energy does not vanish. The energy of this state is

$$E_o = \sum_{\mathbf{k},s}^{\infty} \hbar\omega_{\mathbf{k}} \left(\frac{1}{2}\right) \tag{9.53}$$

which is a divergent quantity (equal to infinity). It is the sum of the energies of the ground state of each harmonic oscillator and it is infinite because there is no upper boundary to the field's frequency even if the volume is finite. There is no satisfactory explanation for the physical significance of this, but, fortunately, we can eliminate this term by shifting the energy of the ground state. This infinite energy shift cannot be detected experimentally, since the experiments measure only energy differences from the ground state of H_{EM}. This infinite constant term, called the zero point energy of the radiation field, is responsible for many interesting effects. Some phenomena which could not be explained using classical fields, such as the Casimir effect, the Lamb shift and spontaneous emission, can be understood as effects caused by the quantum nature of radiation. Spontaneous emission can be explained as stimulated emission by the zero point energy. The attractive force between two parallel, infinite, conducting plates in vacuum, known as the Casimir effect, is produced by the vacuum modes between the plates. Furthermore, according to the semi-classical theory, the $2S_{\frac{1}{2}}$ and $2P_{\frac{1}{2}}$ levels of the hydrogen atom should have the same energy. Some corrections can be made if the interaction between the atomic electron and the vacuum field are taken into account, which causes the energy of the $2S_{\frac{1}{2}}$ level to be $1057MHz$ higher than the $2P_{\frac{1}{2}}$ level. This is called the Lamb shift.

It is possible to calculate the commutator between the field's energy and the electric field using the commutation relations. The commutator does not vanish, so the uncertainty relations say that for a state of definite energy, the electric field must have dispersion. The same thing is true for the magnetic field. This means that in the vacuum state the electric and magnetic fields fluctuate. These vacuum fluctuations are one of the most striking consequences of the quantization of the radiation field. The fluctuations are proportional to the square of the fields because the expectation value of the fields vanish for the vacuum state,[1] so

$$\langle 0|\bar{E}^2|0\rangle = \frac{\hbar}{\epsilon_0 V}\sum_{\mathbf{k}}^{\infty}\omega_{\mathbf{k}} = \frac{\hbar}{\epsilon_0(2\pi)^3}\int_0^{\infty}\omega_{\mathbf{k}}d^3k \qquad (9.54)$$

We see that the fluctuations are infinite for an unbounded set of modes, but in practice, measurements are made over a finite region of time, frequencies and space. Detectors are sensitive to a finite number of frequencies and a measurement is the average over a space–time region and bandwidth. The fluctuations are then finite.

So, there is energy in the light field even when there are no photons around. Not only is there energy, but an infinite amount of it. The most spectacular consequence of this is that two parallel conducting plates would attract each other with a force that arises due to the quantum vacuum. Let's try to estimate the magnitude of this force — known as the Casimir force — using a simple heuristic argument. The force is there because the energy between the plates is smaller when they are closer together, so they accelerate towards each other. The force is equal to the change in energy divided by the change in distance. The trouble is that the energy between the plates is infinite at all times since there are infinitely many modes in between them no matter what the gap is, $U = \sum_k \frac{1}{2}\hbar\omega_k = \infty$. Let's sidestep this uncomfortable issue by pretending that there is an upper cut-off frequency (or, equivalently, wave vector). After all, wave vectors beyond 1/(atomic dimensions) don't make much sense when it comes to the field boundary conditions anyway. Thus, if L^2 is the plate area and their separation is x, we have

$$U \approx L^2 x \int_{1/x}^{1/a_0} \hbar c k k^2 dk = \frac{1}{4}L^2\hbar c\left(\frac{x}{a_0^4} - \frac{1}{x^3}\right) \qquad (9.55)$$

where a_0 is the Bohr radius. The force per unit area is now

Casimir's Force

$$F = -\frac{1}{L^2}\frac{dU}{dx} \sim \frac{1}{x^4} \qquad (9.56)$$

This is the dependence derived by Casimir and the right coefficient is $\pi\hbar c/480$.

An interesting feature of this effect is that it depends crucially on the geometry of the problem. If instead of two parallel plates we had two concentric (and conducting) spheres, they would actually repel each other. We will see that this has an interesting

---

[1]This is because $\hat{a}|0\rangle = \langle 0|\hat{a}^\dagger = 0$ and $E \propto \hat{a} + \hat{a}^\dagger$.

consequence when it comes to electronic energies (see the section on the Lamb shift).

There is an interesting maritime analogy with the Casimir effect that shows just how clearly the wavelike nature of particles **Maritime** is responsible for the somewhat counterintuitive vacuum effects. **Analogy** This is a real-life story about the warning issued to sailors (a few centuries or so ago) that two ships which are docked in a harbor parallel to each other may start to approach each other (and may ultimately crash into each other) due to their small oscillations up and down in the water. The explanation of this effect, which is remarkably similar to the Casimir force, lies in the fact that the waves that exist between the two ships actually cancel each other out, while the waves going away from the ships don't. This means that due to the conservation of momentum the ships recoil towards each other and this is the origin of the maritime force. The force can even be calculated to be

$$F \propto \frac{mhA^2}{T^2} \qquad (9.57)$$

where m is the mass of the ship, h its amplitude of oscillations, A the angle from the normal to the water surface and T the period of oscillations. If $m = 700$ tons, $h = 1.5$ meters, $A = 8$ degrees and $T = 8$ seconds, we obtain the force $F = 10^2$ Newtons, which is by no means a negligible force.

9.5   Coherent States

Coherent states are very important in quantum physics and, in particular, quantum optics. They are the most appropriate quantum representation of what we might call classical light (there is, of course, no such thing as classical light, but coherent states with a large number of photons are closest to what we imagine classical light should be). Before we present these states let's first talk about photon counting statistics. This will provide us with a clue as to how to construct coherent states. Suppose that you sit in front of a laser and register photons as they come out. What is the probability of registering a count in a small interval dt going to be? We'd expect it to be proportional to the intensity of light, $I(t)$, as well as the size of dt (for small time intervals this is true), so that

$$dp(t) = cI(t)dt \qquad (9.58)$$

The coefficient c is there to account for the imperfections of detection and other approximations that might exist. From here we can conclude that the probability for no photon counts in the same interval is

$$1 - dp(t) = 1 - cI(t)dt \tag{9.59}$$

For simplicity we'd like to assume that there are no fluctuations in the intensity during the time we measure. This is of course not true, but it is true for classical light to a good approximation as we will see later on. So, the probability that there are no counts in the interval $(t, t+T)$ is

$$p_0(t, t+T) = \prod (1 - dp(t)) = \left(1 - \frac{cIT}{k}\right)^k \to e^{-cIT} \tag{9.60}$$

This is true only if the counts are truly independent from one time interval to the next, which again is a good approximation. How about the probability for n photon counts in the same interval? Let's introduce the shorthand notation $\mu = cIT$. Then,

Poisson Distribution

$$p_n(t, t+T) = \frac{\mu^n}{n!}e^{-\mu} \tag{9.61}$$

The probability for n counts is μ^n. Why do we divide by "$n!$"? Because this is the number of possible arrangements of counts, and we don't want to over-count them. Therefore the distribution of counts is Poissonian.

Now I would like to argue that the coherent states are the best quantum description of laser light. Why? Coherent states are in some sense closest to the classical states. Here is how. The coherent state of amplitude α is defined by

Coherent State

$$|\alpha\rangle = e^{-|\alpha|^2/2} \sum_n \frac{\alpha^n}{\sqrt{n!}}|n\rangle \tag{9.62}$$

The amplitude for having n photons is given by

$$\langle n|\alpha\rangle = e^{-|\alpha|^2/2} \frac{\alpha^n}{\sqrt{n!}} \tag{9.63}$$

and the probability follows the familiar Poisson distribution

$$p(n) = |\langle n|\alpha\rangle|^2 = e^{-|\alpha|^2} \frac{|\alpha|^{2n}}{n!} \tag{9.64}$$

as we agreed it should. The mean number of photons is given by

$$\langle n\rangle = \sum np(n) = |\alpha|^2 = \langle \alpha|a^\dagger a|\alpha\rangle \tag{9.65}$$

Why did we say that this is the closest quantum representation of the classical laser field? First, because it is possible to absorb the photons without changing the state, i.e. the coherent state is an eigenstate of the annihilation operator

$$a|\alpha\rangle = \alpha|\alpha\rangle \qquad (9.66)$$

How about the time evolution of the coherent state under the Hamiltonian $H = \hbar\omega(a^\dagger a + 1/2)$? It is given by

$$|\psi(t)\rangle = e^{-i\omega t/2}|\alpha e^{-i\omega t}\rangle \qquad (9.67)$$

(solve the free evolution Schrödinger equation to obtain the above expression). So, the state continuously evolves through coherent states, at the frequency ω — another analogy with the classical field. Finally, let's compute the averages of the electric and magnetic fields. They are (prove this by using the expansion of coherent states in terms of number states and then using your knowledge of the action of creation and annihilation operators on number states)

$$\langle\alpha(t)|q|\alpha(t)\rangle = \left(\frac{2\hbar}{\omega}\right)^{1/2}|\alpha|\cos(\omega t - arg\alpha) \qquad (9.68)$$

$$\langle\alpha(t)|p|\alpha(t)\rangle = -(2\hbar\omega)^{1/2}|\alpha|\sin(\omega t - arg\alpha) \qquad (9.69)$$

and so they oscillate the way that the classical electric and magnetic fields do.

Coherent states are also similar to classical states in the sense that they are also minimum uncertainty states (although this, of course, has no classical analogue). To be able to calculate this, we need to remember that the position of an oscillator is given by

$$\hat{q} = (\hbar/2\omega)^{1/2}(\hat{a} + \hat{a}^\dagger) \qquad (9.70)$$

Thus

$$\hat{q}^2 = \frac{\hbar}{2\omega}[\hat{a}^2 + (\hat{a}^\dagger)^2 + 2\hat{a}^\dagger\hat{a} + 1] \qquad (9.71)$$

and

$$\langle v|\hat{q}^2|v\rangle = \frac{\hbar}{2\omega}[v^2 + (v^*)^2 + 2|v|^2 + 1] \qquad (9.72)$$

The dispersion that will feature in the Heisenberg uncertainty relation is

$$\langle v|\Delta\hat{q}^2|v\rangle = \langle v|(\hat{q}^2 - \langle v|\hat{q}|v\rangle)^2|v\rangle = \frac{\hbar}{2\omega} \qquad (9.73)$$

We can similarly show that

$$\langle v|\Delta\hat{p}^2|v\rangle = \frac{\hbar\omega}{2} \qquad (9.74)$$

The product of uncertainties is

$$\sqrt{\langle v|\Delta\hat{q}^2|v\rangle}\sqrt{\langle v|\Delta\hat{p}^2|v\rangle} = \frac{\hbar}{2} \qquad (9.75)$$

What is the phase of the coherent state? The simple answer is $\phi = \arg(\alpha)$. But why? Surely, the phase is something we can measure experimentally and therefore there should be a Hermitian operator associated with it.[2] Let's call this operator $\hat{\Phi}$, so that we would like to have that

$$\langle\alpha|\hat{\Phi}|\alpha\rangle = \arg(\alpha) \qquad (9.76)$$

How do we define $\hat{\Phi}$, so that we reproduce the above result? We have to first define a different basis to the number states basis

$$|\phi_j\rangle := \frac{1}{\sqrt{N}}\sum_{n=0}^{N-1} e^{2\pi ijn/N}|n\rangle \qquad (9.77)$$

These states are orthogonal to each other, i.e. $\langle\phi_i|\phi_j\rangle = \delta_{ij}$ (prove it). Now suppose that we define the phase operator to be (in terms of its eigenvectors and eigenvalues)

$$\hat{\Phi} := \sum_j \left(\frac{2\pi j}{N}\right)|\phi_j\rangle\langle\phi_j| \qquad (9.78)$$

In order to compute the phase of the coherent state we have to calculate $\langle\alpha|\hat{\Phi}|\alpha\rangle$. We need to know that

$$\langle n|\phi_j\rangle = \frac{1}{\sqrt{N}}e^{2\pi ijn/N} \qquad (9.79)$$

which follows straight from the definition in equation (9.77). So,

$$\langle\alpha|\hat{\Phi}|\alpha\rangle = e^{-|\alpha|^2}\sum_{mnj}\frac{\alpha^{*n}\alpha^m}{\sqrt{n!m!}}\frac{(2\pi j)}{N}\langle n|\phi_j\rangle\langle\phi_j|m\rangle \qquad (9.80)$$

$$= e^{-|\alpha|^2}\sum_{mnj}\frac{\alpha^{*n}\alpha^m}{\sqrt{n!m!}}\frac{(2\pi j)}{N}\frac{1}{N}e^{2\pi ij(n-m)/N} \qquad (9.81)$$

Given that $\alpha = re^{i\phi}$ (any complex number can be written this way), we have that

$$\langle\alpha|\hat{\Phi}|\alpha\rangle = \sum_j \frac{2\pi j}{N}(\frac{1}{N} + 2\frac{e^{-r^2}}{N}\sum_{n>m}\frac{r^n r^m}{\sqrt{n!m!}}\cos(2\pi ij(n-m)/N))$$

$$(9.82)$$

---

[2]The approach I follow here is from Pegg and Barnett, Europhys. Lett. **6**, 483 (1998). But there are many others.

For large N we can convert the last expression into an integral and evaluating the integral we obtain the desired solution that the expectation value of the phase operator is indeed the phase (argument) of the coherent state (amplitude). See if you can work this out — but don't despair if you don't succeed (have a look at the paper by Pegg and Barnett cited in Footnote 2).

9.6 Composite Systems — Tensor Product Spaces

When we have more than one system to describe quantum mechanically, for example two atoms, then we have to use a mathematical structure called a tensor (or direct) product of Hilbert spaces. Each one of those systems will live in its own Hilbert space, but when we wish to describe them collectively, we need to combine their Hilbert spaces in some specific way. This construct, called a tensor product, is a very simple beast. Suppose that we have two two level atoms, A and B. Their states are $|g\rangle_A$ and $|e\rangle_A$ for the atom A, and $|g\rangle_B$ and $|e\rangle_B$ for the atom B. What are the states of the joint system? Well, there are four possibilities: both atoms are in the ground state, A is in the ground state and B in the excited state, vice versa, and both atoms are in the excited state. Mathematically, we label these as

$$|g\rangle_A \otimes |g\rangle_B, |g\rangle_A \otimes |e\rangle_B, |e\rangle_A \otimes |g\rangle_B, |e\rangle_A \otimes |e\rangle_B \qquad (9.83)$$

The $\otimes$ sign is there to signify the existence of two independent systems, which together live in a Hilbert space that is a tensor product of the two individual Hilbert spaces. We will frequently omit the subscripts A and B and write

$$|g, g\rangle \qquad (9.84)$$

instead of $|g\rangle_A \otimes |g\rangle_B$, and the first ket will always be assumed to belong to A while the second ket belongs to B. Now if there is an operator acting on A, say $\hat{O}_A$, and another one acting on B, say $\hat{O}_B$, then their joint action, labelled as $\hat{O}_A \otimes \hat{O}_B$, is defined as

$$\hat{O}_A \otimes \hat{O}_B(|g\rangle_A \otimes |g\rangle_B) := (\hat{O}_A|g\rangle_A) \otimes (\hat{O}_B|g\rangle_B) \qquad (9.85)$$

This is the key property of the tensor product structure, i.e. the systems preserve their individuality, so that operators from A don't mix with operators from B and vice versa. Therefore, we can have an operator that excites the atom A, i.e. it performs the operation $\hat{\sigma}_+^A|g\rangle_A = |e\rangle_A$, without affecting the second atom. This action is described as

$$(\hat{\sigma}_+^A \otimes I)(|g\rangle_A \otimes |g\rangle_B) = \hat{\sigma}_+^A |g\rangle_A \otimes I |g\rangle_B \qquad (9.86)$$

$$= |e\rangle_A \otimes |g\rangle_B \qquad (9.87)$$

Note that the systems behave completely independently, so that $(A \otimes B)^\dagger = A^\dagger \otimes B^\dagger$ (unlike when the operators act on the same system, when we would have $(AB)^\dagger = B^\dagger A^\dagger$).

The two systems can, of course, always interact. In that case, the notation remains the same as the systems are still different, it's just that their states become dependent on each other, i.e. correlated.[3] The notion of the tensor product of two Hilbert spaces is nicely illustrated by the operation of the beam splitter. We discuss this subject next.

9.6.1 *Beam splitters*

When we talk about a beam splitter we need to invoke four different quantum systems: two input light field modes (harmonic oscillators) and two output light field modes (also harmonic oscillators). Let's call the modes a, b, c and d respectively. We now want to describe in full detail the situation we analyzed at the beginning when we talked about the Mach–Zehnder interferometer. Imagine that mode a contains one photon and mode b contains no photons initially. The state is

$$|1\rangle_a \otimes |0\rangle_b \qquad (9.88)$$

---

[3]The appropriate terminology here is that the states of the system will become entangled. Entanglement is a term coined by Schrödinger [Schrodinger (1935)] to describe an excess of correlations existing between quantum mechanical systems (photons, atoms, etc.) that can in no way be reproduced using the standard (Newtonian) classical physics. Although these correlations have been used to highlight a number of (apparent) paradoxes at the heart of quantum physics, they have nevertheless been confirmed in a number of different experiments since the beginning of the eighties.

In these experiments, the outcomes of independent measurements on two quantum systems are shown to be so highly correlated that no classical (or, more precisely — local realistic) model can ever mimic them. A high degree of correlation here means that once the state of one of the two systems is known, we immediately know the state of the other system. These correlations have been under a great deal of scrutiny since the birth of quantum theory and have more recently been linked to the speed-up in a quantum computer. I will use them at the end to illustrate the protocol known as quantum teleportation. I would highly recommend reading the collection of John Bell's papers entitled *Speakable and Unspeakable in Quantum Mechanics* (Cambridge University Press, 1990).

What is the state after the "single photon" has passed through the beam splitter?[4] It is

$$\frac{1}{\sqrt{2}}(|1\rangle_c \otimes |0\rangle_d + |0\rangle_c \otimes |1\rangle_d) \qquad (9.89)$$

This state is an equal superposition of a single photon in mode c and a single photon in the mode b. Let's see what happens if we have the number state as an input in one mode and a vacuum state in the other mode. Then we can write

$$|n\rangle|0\rangle = \frac{(a^\dagger)^n}{\sqrt{n}}|0\rangle|0\rangle \qquad (9.90)$$

Rather then performing any transformations let's see if we can guess what happens. First of all it could happen that all n photons go to mode c. Then $n-1$ may go to c and one photon then goes to d, and so on until the final possibility when all n photons go to mode d. Finally, all of these possibilities happen simultaneously, so that the resulting state at the output of the beam splitter is

$$\sum_i c_i |i\rangle_c |n-i\rangle_d \qquad (9.91)$$

Now all we have left to do is determine the amplitudes c_i. We know two things:

(1) it's a 50–50 beam splitter and
(2) all the mod squares have to add up to one.

This reminds us very much of the binomial distribution. Remember that

$$(a+b)^n = a^n + \binom{n}{1}a^{n-1}b^1 + \binom{n}{2}a^{n-2}b^2 + \cdots + b^n \qquad (9.92)$$

In our case $a = b = 1/2$ so that the whole sum is unity. Thus, we can guess

$$c_i = \sqrt{\binom{n}{i}2^{-n}} \qquad (9.93)$$

---

[4]I should say that we don't really have any single photon sources at present. This is a purely technological problem and it is not fundamental. In reality we would have a superposition of different number of photons just like we did when we discussed coherent states.

So this is really like coin tossing[5] except that all the possibilities add up into a superposition, rather then being realized separately as would be the case for a classical random process. The only additional element we can have is a phase (which would be either 1 or i depending on whether the photon is reflected not as before. We now generalize this treatment to coherent states.

Exercise Compute what happens when a coherent state goes through a 50–50 beam splitter. Suppose that in one port of the beam splitter, port a, we have the coherent state $|\alpha\rangle$ and in the other port, port b, we have nothing (i.e. the vacuum state $|0\rangle$).

Note that there are two ways we can describe quantum evolution. One is to specify what happens to states, i.e. give the evolution of a set of orthogonal states into another set of orthogonal states. The other way is to specify what happens to the corresponding operators. Let's see how this would work in the case of the beam splitter. The rules for the annihilation (creation) operators at the beam splitter are

$$\hat{a} = (\hat{c} + \hat{d})/\sqrt{2}$$
$$\hat{b} = (\hat{c} - \hat{d})/\sqrt{2} \tag{9.94}$$

where c and d are the two output ports and $\hat{c}$ and $\hat{d}$ the corresponding annihilation operators. The reason for this form of the operators is that the two output states have to lead to two orthogonal states (like $|0\rangle + |1\rangle$ and $|0\rangle - |1\rangle$). We know that the input coherent state can be written as

$$|\alpha\rangle = e^{-|\alpha|^2/2} \sum_n \frac{\alpha^n}{\sqrt{n!}} \frac{(a^\dagger)^n}{n!} |0\rangle \tag{9.95}$$

Thus, after the beam splitter the state is

$$|\text{output}\rangle = e^{-|\alpha|^2/2} \sum_n \frac{\alpha^n}{\sqrt{n!}} \frac{(c^\dagger + d^\dagger)^n}{n!} |0\rangle \tag{9.96}$$

Using the *binomial formula*, and rearranging the terms (prove it!), this becomes

$$|\text{output}\rangle = |\alpha/\sqrt{2}\rangle \otimes |\alpha/\sqrt{2}\rangle \tag{9.97}$$

This shows that each output port contains a coherent state with half the intensity (i.e. one over square root 2 of the amplitude)

[5]Tossing a 50–50 coin n times produces the same kind of binomial statistics and this is because a photon, like a coin, has a binary choice at the beam splitter: left or right. The difference, though, is that the photon, unlike the coin, can and does go both ways.

of the original state. Hence coherent states behave like classical states at the beam splitter; they are split in half.

9.6.2   *Generation of coherent states*

How do we generate a coherent state? Now I want to show you this in a very simple way. Coherent states are obtained by emission from a classical current. By classical I mean that the current will be described by a classical current vector — no need to quantize it. It's just a bunch of electrons shaking up and down. What we present is also a kind of semi-classical picture, but now the field is quantized and the matter is classical — the opposite of what we've had so far. The vector potential interacts with the current in the following way:

$$\hat{V}(t) = \int \mathbf{J}(\mathbf{r}, t) \mathbf{A}(\mathbf{r}, t) d^3 r \qquad (9.98)$$

where

$$A = -i \sum_k \frac{1}{\nu_k} e_k E_k e^{-i\nu_k t + ikr} + \text{h.c.} \qquad (9.99)$$

The Schrödinger equation is now given by

$$\frac{d}{dt}|\psi(t)\rangle = -\frac{i}{\hbar} V |\psi(t)\rangle \qquad (9.100)$$

The vector potential potentials at different times don't commute. So, we cannot integrate the equation the usual to obtain

$$|\psi(t)\rangle = e^{-\frac{i}{\hbar} \int dt V(t)} |\psi(0)\rangle \qquad (9.101)$$

Wrong Way of Integrating

However, due to a set of lucky circumstances the solution is the same as performing this integration "plus" an extra phase which is irrelevant for the state. We can write it as

$$e^{-\frac{i}{\hbar} \int dt V(t)} = \prod_k e^{\alpha_k a_k^\dagger - \alpha^* a_k} \qquad (9.102)$$

where

$$\alpha_k = \frac{1}{\hbar \nu_k} E_k \int dt \int dr e_k J_k e^{-i\nu_k t + ikr} \qquad (9.103)$$

If the initial state is the vacuum state, i.e. $|\psi(0)\rangle = |0\rangle$, then we have

$$|\psi(t)\rangle = \prod_k e^{\alpha_k a_k^\dagger - \alpha^* a_k} |0\rangle \qquad (9.104)$$

The state $|0\rangle$ in the previous equation actually represents a vacuum state in all the modes:

$$|0\rangle = \prod_k |0_k\rangle \qquad (9.105)$$

This results in a number of coherent states, one for each more labelled by **k**:

$$|\psi(t)\rangle = \prod_k |\alpha_k\rangle := |\alpha_1\rangle \otimes |\alpha_2\rangle \cdots |\alpha_N\rangle \qquad (9.106)$$

We are going to be concerned mainly with coherent states in one of those modes, but I wanted to show you the full formalism so that you gain a deeper understanding of this phenomenon.

9.7 Bosonic Nature of Light

There are two types of particles in Nature: bosons and fermions. It is frequently loosely stated that "bosons tend to bunch", while "fermions tend to anti-bunch". Since photons are bosons we have to investigate these statements more closely. We are in a position to do so now more accurately.[6] First of all, *how come that this fundamental aspect has not been needed so far in our treatment?* Well, for most of it we have worked in the classical and semi-classical approximation and there was no need to take into account the bosonic nature of light directly. Now, however, we have quantized the field as well and we need to consider all the quantum features. The reason why we still have been fine in the first part of this section is because we have been looking at the single mode of the field. Now we will consider more than one mode and the typical bosonic effect will manifest itself in the form of bunching of photons in these modes.

Let's now describe a very simple experiment which demonstrates that identical quantum particles are indistinguishable. In particular, we will see that a consequence of this is that photons tend to bunch. Here is the experiment showing this: imagine having two photons at a beam splitter (one in each port as shown in Figure 9.2). The transformation that they would undergo is

$$|\text{up}\rangle \Longrightarrow i|\text{up}\rangle + |\text{down}\rangle \qquad (9.107)$$

$$|\text{down}\rangle \Longrightarrow i|\text{down}\rangle + |\text{up}\rangle \qquad (9.108)$$

---

[6]This is usually not described in courses at this level. I don't really understand why. The concept is not so difficult and it is really of paramount importance for almost any application of quantum physics.

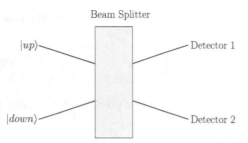

Fig. 9.2 Operation of a beam splitter. The mathematics of the transformations is given in the text.

Suppose that we have photons simultaneously impinging from each side of the beam splitter. Then due to their "mixing" at the beam splitter, the transformation will be (altogether)

$$|\text{up}\rangle|\text{down}\rangle \Longrightarrow (i|\text{up}\rangle + |\text{down}\rangle)(i|\text{down}\rangle + |\text{up}\rangle) \qquad (9.109)$$
$$= i(|\text{down}\rangle|\text{down}\rangle + |\text{up}\rangle|\text{up}\rangle) + (|\text{down}\rangle|\text{up}\rangle - |\text{up}\rangle|\text{down}\rangle) \quad (9.110)$$

Now if there is absolutely no way of distinguishing the photons (not even in principle and not just in practice), then we have to symmetrize the state, i.e. have to impose that

$$|\text{down}\rangle|\text{up}\rangle = |\text{up}\rangle|\text{down}\rangle \qquad (9.111)$$

So, the above state is reduced to

$$|\text{down}\rangle|\text{down}\rangle + |\text{up}\rangle|\text{up}\rangle \qquad (9.112)$$

so, like we said, either both of the photons go up or down. This is the well-known effect of bunching of bosons. (Note: if the particles were fermions, then $|\text{down}\rangle|\text{up}\rangle = -|\text{up}\rangle|\text{down}\rangle$ and the surviving terms above would be the ones where we have one fermion "up" and the other one "down". Therefore, fermions anti-bunch, which is known as the Pauli exclusion principle, i.e. two fermions cannot occupy the same state.)

Let us now imagine a slightly more complex situation: suppose that each photon has a polarization degree of freedom (as they do). What would happen at a beam splitter if they had different polarizations — one "horizontal" and the other one "vertical". Well, they are then distinguishable in principle and no "funny" bunching would occur. What we would register is all four possibilities, i.e. both "up", both "down", the vertical photon "up" and horizontal "down", and vice versa, all of which would be equally likely. On the other hand, if the polarizations are the same, then the bunching only occurs again.

Let us finally complicate the matter even further. Let the photons have different polarizations initially, but just before they encounter the beam splitter, transform the polarization of one of the photons so that it is identical to the other one. Then, they become indistinguishable and again we have coincidence. The experiment described is by Ghosh, Hong, Ou and Mandel (see, for example, Phys. Rev. A **34**, 3962 (1986)).

Where does the effect of (quantum) particle statistics come from? Remember the postulates of quantum mechanics?

(1) States of physical systems are vectors in a Hilbert space;
(2) observables are Hermitian operators acting on the Hilbert space;
(3) measurements give probabilistic outcomes according to the usual trace rule;
(4) when there is no measurement, the system obeys the Schrödinger equation.

These are the postulates. Notice that there is no mention of particle statistics. At this level, particle statistics and the assumption about the fermions and bosons is an *ad hoc* hypothesis, which is verified by experiment. So, there is no way of deriving it from the elementary quantum postulates. We need something more — the full quantum field theory, i.e. relativity appears to be crucial. But this is well beyond our intended level here.

9.8 Polarization: The Quantum Description

Polarization should just be another quantum observable and it turns out that it is. So, there are really two parts to the state of light: there is the "spatial" part describing the state as a superposition of various photon number states, but there is also the polarization, mathematically assigned another Dirac ket vector. So, in general the full state of light should be written as

$$|n\rangle_H \otimes |m\rangle_V \qquad (9.113)$$

Here the first ket refers to the number of photons with horizontal polarization and the second ket represents the number of photons with vertical polarization. So, what is the true description of the sea of light that surrounds us all? We have many different frequencies, $\omega_1, \omega_2, \ldots$ and each one has two harmonic oscillators associated with it. This can be mathematically expressed as

$$|\text{Field}\rangle = (|n_1\rangle_H \otimes |m_1\rangle_V)_{\omega_1}$$
$$\otimes (|n_2\rangle_H \otimes |m_2\rangle_V)_{\omega_2}$$
$$\otimes (|n_3\rangle_H \otimes |m\rangle_V)_{\omega_3} \cdots \qquad (9.114)$$

This is, in fact, the state of a cavity full of radiation. Notice how Planck's and Einstein's treatment look simplistic compared to this!

Suppose that we have a single photon at frequency ω and polarized at $45°$. How can we represent this? The logic is that $45°$ is an equal superposition of the horizontal and vertical "photon", so that

$$|\Psi_{45}\rangle = \frac{1}{\sqrt{2}}(|1\rangle_H \otimes |0\rangle_V + |0\rangle_H \otimes |1\rangle_V) \qquad (9.115)$$

and we indeed have either one horizontal photon and no vertical photon or vice versa. To make our life easier, I will use a shorthand notation to represent the above:

$$|\Psi_{45}\rangle = \frac{1}{\sqrt{2}}(|H\rangle + |V\rangle) \qquad (9.116)$$

Armed with this understanding we attack the notion of unpolarized light (the most common around us — lamps, candles, sun's light etc. — are all really unpolarized).

9.8.1 *Unpolarized light — mixed states*

Suppose that light is polarized at $45°$ and we detect how many photons are polarized in the x direction and how many are in the y direction. Say it's 50–50, half of them will be x polarized and the rest will be y polarized. Good. Now suppose light is unpolarized in which case the polarization is completely random. Since every photon is polarized in a random direction, we would also record half of the photons in the x direction and half of them in the y direction. Therefore, it is tempting to conclude that there is no difference between the unpolarized light and linearly polarized light. But this is completely wrong!

Let's see how to represent completely unpolarized light. First of all the light polarized at $45°$ is represented by

$$|\Psi_{45}\rangle = \frac{1}{\sqrt{2}}(|H\rangle + |V\rangle) \qquad (9.117)$$

This is an equal superposition of the horizontal and vertical polarization (and it is therefore obvious why it is a 45 degree polarization). This state indeed yields one measurement of H, V polarization: half probability of the horizontal polarization and half probability of the vertical polarization. On the other hand, the unpolarized light is written as

$$\rho_{\text{unpol}} = \frac{1}{2}(|V\rangle\langle V| + |H\rangle\langle H|) \tag{9.118}$$

The same is true for the state $|H\rangle$. This is known as a mixed state. Note that the way we wrote a mixed state is like an operator, and not like the 45 degree polarization pure state. What about the probability of obtaining a state $|V\rangle$? It is

$$p_V := \langle V|\rho|V\rangle = \frac{1}{2} \tag{9.119}$$

And in terms of preparation this state is obtained by preparing each photon randomly in the state $|H\rangle$ or $|V\rangle$. For example we generate a photon and then toss a coin. If it is heads we prepare the photon in the state $|H\rangle$, otherwise we prepare $|V\rangle$. So how can we distinguish the pure state $|H\rangle + |V\rangle$ from the mixed state? Well, we just have to perform a measurement in a different basis. For example, we can use the basis

$$|+45\rangle = \frac{1}{\sqrt{2}}(H\rangle + |V\rangle) \tag{9.120}$$

$$|-45\rangle = \frac{1}{\sqrt{2}}(H\rangle - |V\rangle) \tag{9.121}$$

Exercise Calculate the probabilities for the two states if the photon is in the state $|H\rangle + |V\rangle$; or if the photon is in ρ_{unpol}.

Therefore we conclude that there is a big (*observable*) difference between the two states. There is much more to mixed states than I have discussed here. They become particularly important in applications of quantum mechanics to real-life situations where the system under investigation interacts with its (possibly large and intractable) environment. It is this interaction with the environment that makes states of the system mixed. If we think of the state of the system as the collection of all information that we have about that system, then having all the information implies that the state of the system is pure. However, if the environment gains some information about the system "before" we manage to, then we are left with less then maximal knowledge of the state of the system and this means that we have a mixed state. Quite a large proportion of research time is nowadays devoted to studying mixed states, but we unfortunately have no more space for it here.

Chapter 10

Interaction of Light with Matter III

We have already indicated that light and matter interact by exchanging excitations much as two pendulums would do if they were coupled. We now turn to the fully quantized treatment which was originally due to Dirac. He realized that the appropriate way of treating this problem was to have a Hamiltonian consisting of three parts: the free atom Hamiltonian, the free field Hamiltonian and the Hamiltonian describing their joint interaction. The key point is to understand the interaction Hamiltonian. There are, of course, many different Hamiltonians which could describe the interaction, and there are many different ways for these two systems to interact. We will investigate the most common Hamiltonian: the Jaynes–Cummings Hamiltonian [Jaynes and Cummings (1963)].

The point that we have now reached is sometimes the starting point of more advanced quantum optics treatments such as, for example, [Scully (1997)].

10.1 Fully Quantized Treatment

As was discussed above, the total atom field Hamiltonian can be written as

$$H = H_A + H_F + H_I \qquad (10.1)$$

and as we know H_A and H_F are well-defined once we know the transition frequency of the two level atom, and the frequency of the harmonic oscillator respectively. The interaction Hamiltonian H_I, on the other hand, is somewhat arbitrary. Let's have a look at the semi-classical Hamiltonian we have used before

$$H_I = \mu B \sigma_z \qquad (10.2)$$

This represents a two level system with energy levels $\pm \mu B$. The field here is represented by a real number B which tells us about its strength. But there is nothing quantum mechanical about it. Quantum mechanically it should be represented by the creation and annihilation operators. Thus the full treatment must quantize the field and then add its energy to the Hamiltonian as well as the term that describes how it interacts with the two level system. We turn to this problem next.

10.2   Jaynes–Cummings Model

The Jaynes–Cummings model (JCM) is a fundamental model widely used in quantum optics. Although at first sight it appears to be very simple, the model has been studied for more than 30 years. New, exciting and surprising discoveries are still being made. The JCM is the first fully quantized model of the interaction between a two level atom (or indeed any other two level system) and a quantized, monochromatic electromagnetic field. In order to present this model, we shall briefly remind ourselves of the process of quantization of the electromagnetic field, but from a slightly different and more appropriate direction.

The time development of an isolated quantum system is completely determined once we specify the corresponding Hamiltonian. Let us, therefore, represent the states of the atom by vectors $|g\rangle$ (ground) and $|e\rangle$ (excited). The field is at the same time represented by the energy eigenvectors of a quantum mechanical harmonic oscillator, i.e. the number states $|n\rangle$. Invoking the action of raising and lowering operators

$$a^\dagger|n\rangle = \sqrt{n+1}|n+1\rangle \tag{10.3}$$

$$a|n\rangle = \sqrt{n}|n-1\rangle \tag{10.4}$$

and taking the energy of the ground state to be zero, and that of the excited state to be $\hbar\omega_0$, we can write the Hamiltonian of the composite atom field system as

$$H = \hbar\omega_0|e\rangle\langle e| + a^\dagger a\hbar\omega + V \tag{10.5}$$

where we omitted the "vacuum field" zero point energy term from the field. The interacting Hamiltonian, V, is given by the dipole field interaction energy:

$$V = -e\mathbf{r}\mathbf{E} = -exE_x \tag{10.6}$$

where e is the electric unit charge and is not to be confused with the excited state, $|e\rangle$. Taking the rotating wave approximation, i.e.

ignoring the rapidly varying terms, we get the following Hamiltonian:

$$H = \hbar\omega_0|e\rangle\langle e| + a^\dagger a\hbar\omega + \hbar\lambda(a^\dagger\sigma_- + a\sigma_+) \qquad (10.7)$$

Jaynes–Cummings Hamiltonian

known as the Jaynes–Cummings Hamiltonian. Here

$$\lambda = -\frac{\langle e|ex|g\rangle\sqrt{\frac{\hbar\omega}{\epsilon_0 L}}\sin kz}{\hbar} \qquad (10.8)$$

is the Rabi frequency from before (just $dE/\hbar$, where E is obtained from equation (9.46)). Note that the interacting term has a very simple and natural interpretation: the first interaction term indicates that the atom is de-excited and the field receives this quantum of excitation, while the second term is the exact reverse. In general, after some time t, the atom and the field are in an entangled state of the form

$$|\Psi(t)\rangle = \sum_{n=0}^{\infty}(a_n(t)|g\rangle \otimes |n\rangle + b_n(t)|e\rangle \otimes |n\rangle) \qquad (10.9)$$

Solving Schrödinger's equation we obtain

$$a_n(t) = ac_n\cos\lambda\sqrt{n}t - ibc_{n-1}\sin\lambda\sqrt{n}t \qquad (10.10)$$
$$b_n(t) = bc_n\cos\lambda\sqrt{n+1}t - iac_{n+1}\sin\lambda\sqrt{n+1}t \qquad (10.11)$$

where we have assumed that the atom and the field are initially in a disentangled state of the form

$$|\Psi(t=0)\rangle = (a|g\rangle + b|e\rangle) \otimes \sum_{n=0}^{\infty}c_n|n\rangle \qquad (10.12)$$

We will use the evolution equations (10.10 and 10.11) in the following section. See if you can obtain this solution; if not, you need to be patient for a little while to see how this is done.

What happens if we have more than one mode? Then we have to add all this information into the Hamiltonian. Let us now write down the complete description to see how to arrive at the Jaynes–Cummings model more rigorously. The Hamiltonian for an atom with a single electron interacting with a field in the dipole approximation is

$$\hat{H} = \hat{H}_a + \hat{H}_{EM} - e\hat{\mathbf{r}} \cdot \hat{\mathbf{E}} \qquad (10.13)$$

where $\hat{H}_a$ and $\hat{H}_{EM}$ are the Hamiltonian for the unperturbed atom and free field, respectively. The vector $\mathbf{r}$ is the position of the electron. $\hat{H}_{EM}$ is given by the usual harmonic oscillator Hamiltonian and we find H_o from the equation (8.19),

$$\hat{H}_a = \sum_i \hbar\omega_i |i\rangle\langle i| \qquad (10.14)$$

where $\{|i\rangle\}$ is the complete set of the atomic energy levels. Now, we can express $e\mathbf{r}$ in terms of the $\{|i\rangle\}$ states using the usual matrix form for the dipole

$$e\mathbf{r} = \sum_i \wp_{i,j} |i\rangle\langle j| \qquad (10.15)$$

The electric field for the atom at the origin is given by

$$\hat{\mathbf{E}} = \sum_{\mathbf{k}} \bar{\epsilon}_{\mathbf{k}} E_{\mathbf{k}} (\hat{a}_{\mathbf{k}} + \hat{a}_{\mathbf{k}}^\dagger) \qquad (10.16)$$

where $E_{\mathbf{k}} = (\hbar\omega_{\mathbf{k}}/2\epsilon_o V)^{1/2}$. Defining the coupling constant as

$$g_{\mathbf{k}}^{ij} = -\frac{\wp_{ij} \cdot \bar{\epsilon}_{\mathbf{k}} E_{\mathbf{k}}}{\hbar} \qquad (10.17)$$

the Hamiltonian can be expressed as

$$\hat{H} = \sum_i \hbar\omega_i |i\rangle\langle i| + \sum_{\mathbf{k},s} \hbar\omega_{\mathbf{k}} \left(\hat{a}_{\mathbf{k}s}^\dagger \hat{a}_{\mathbf{k}s} \right) + \hbar \sum_{\mathbf{k},i,j} g_{\mathbf{k}}^{ij} (\hat{a}_{\mathbf{k}} + \hat{a}_{\mathbf{k}}^\dagger) |i\rangle\langle j| \qquad (10.18)$$

This Hamiltonian involves a number of different modes (frequencies) of the radiation field. Trying to solve the Schrödinger equation with this Hamiltonian would be a difficult task indeed. Therefore we will simplify the situation (without making it unrealistic). Considering only one mode of the field with frequency ν and a two level atom, i.e $i = a, b$ and $g_{\mathbf{k}}^{ab} = g_{\mathbf{k}}^{ba} = g$ (because of $\wp_{ab} = \wp_{ba}$), the Hamiltonian takes the form

$$\hat{H} = \hbar\omega_a |a\rangle\langle a| + \hbar\omega_b |b\rangle\langle b| + \hbar\nu \left(\hat{a}^\dagger \hat{a} \right) + \hbar g(\hat{a} + \hat{a}^\dagger)(|a\rangle\langle b| + |b\rangle\langle a|) \qquad (10.19)$$

Using the completeness relation for the atom levels, defining $\omega = \omega_b - \omega_a$ and dropping the constant term, we can rewrite the Hamiltonian in the following way:

$$\hat{H} = \frac{\hbar\omega}{2}\hat{\sigma}_z + \hbar\nu \left(\hat{a}^\dagger \hat{a} \right) + \hbar g(\hat{a} + \hat{a}^\dagger)(\hat{\sigma}_+ + \hat{\sigma}_-) \qquad (10.20)$$

where

$$\hat{\sigma}_z = |a\rangle\langle a| - |b\rangle\langle b| \qquad (10.21)$$

$$\hat{\sigma}_+ = |a\rangle\langle b| \qquad (10.22)$$

$$\hat{\sigma}_- = |b\rangle\langle a| \qquad (10.23)$$

It can been seen that the sigma operators defined above satisfy the same algebra as Pauli matrices so the problem of a spin half particle in a magnetic field is mathematically equivalent to the problem studied here (Rabi's problem as it were). The energy term $\hat{a}^\dagger \hat{\sigma}_+$ corresponds to a atom which undergoes a transition form the excited level to the ground state and emits a photon, and $\hat{a}\hat{\sigma}_-$ describes a atom that absorbs a photon and is excited from the ground state to the excited state. Both processes conserve energy. The other two terms in the interaction energy are non-conservative terms so they must be dropped out from the Hamiltonian. This is the rotating wave approximation. So finally the JCM Hamiltonian is

$$\hat{H} = \hbar\omega\hat{\sigma}_z + \hbar\nu\left(\hat{a}^\dagger\hat{a}\right) + \hbar g(\hat{a}\hat{\sigma}_- + \hat{a}^\dagger\hat{\sigma}_+) \qquad (10.24)$$

$$H_o = \hbar\omega\hat{\sigma}_z + \hbar\nu\left(\hat{a}^\dagger\hat{a}\right) \qquad (10.25)$$

$$H_I = \hbar g(\hat{a}\hat{\sigma}_- + \hat{a}^\dagger\hat{\sigma}_+) \qquad (10.26)$$

In the interaction picture the interaction Hamiltonian is

$$\gamma = e^{\frac{iH_o t}{\hbar}} H_I e^{-\frac{iH_o t}{\hbar}} = \hbar g(\hat{\sigma}_+\hat{a}e^{i\Delta t} + \hat{\sigma}_-\hat{a}^\dagger e^{-i\Delta t}) \qquad (10.27)$$

where $\Delta = \omega - \nu$ is the detuning, the usual difference between the driving and natural frequency. The wave function will be a linear combination of states for which the atom is in level $|a\rangle$ and n photons in the field, i.e., $|a\rangle \otimes |n\rangle = |a,n\rangle$ and states in which the atom is in level $|b\rangle$ and n photons in the field

$$|\psi(t)\rangle = \sum_n C_{a,n}(t)|a,n\rangle + C_{b,n}(t)|b,n\rangle \qquad (10.28)$$

$C_{a,n}(t)$ and $C_{b,n}(t)$ are time-varying probability amplitudes. The equation of motion for the wave function is

$$\frac{\partial}{\partial t}|\psi(t)\rangle = -\frac{i}{\hbar}\gamma|\psi(t)\rangle \qquad (10.29)$$

The interaction causes transitions between $|a,n\rangle$ and $|b,n+1\rangle$ only, so the equations of motion for the probability amplitudes are

$$\dot{C}_{a,n} = -ig\sqrt{n+1}e^{i\Delta t}C_{b,n+1} \qquad (10.30)$$

$$\dot{C}_{b,n+1} = -ig\sqrt{n+1}e^{-i\Delta t}C_{a,n} \qquad (10.31)$$

We have an infinite set of decoupled Schrödinger equations, each pair identified by the number of photons that are present when the atom is in the ground state. The solution to the equation when the atom initial state is $|a\rangle$, i.e., $C_{a,n}(0) = C_n(0)$ and $C_{b,n+1}(0) = 0$ is

$$C_{a,n}(t) = C_n(0) \left[\cos\left(\frac{\Omega_n t}{2}\right) - \frac{i\Delta}{\Omega_n} \sin\left(\frac{\Omega_n t}{2}\right) \right] e^{i\Delta t/2} \quad (10.32)$$

$$C_{b,n+1}(t) = -C_n(0) \frac{2ig\sqrt{n+1}}{\Omega_n} \sin\left(\frac{\Omega_n t}{2}\right) e^{-i\Delta t/2} \quad (10.33)$$

Where the Rabi frequency is $\Omega_n = \sqrt{\Delta^2 + 4g^2(n+1)}$. We now calculate the inversion and find

$$W(t) = \sum_n |C_n(0)|^2 \left[\frac{\Delta^2}{\Omega_n^2} + \frac{4g^2(n+1)}{\Omega_n^2} \cos(\Omega_n t) \right] \quad (10.34)$$

The system oscillates with Rabi frequency Ω_n as expected even in the initial vacuum field $|C_n(0)|^2 = 0$. This will be relevant in the next section when we talk about spontaneous emission. A proper model should take into account the infinite modes of the vacuum field which is exactly what we will do.

What happens if the initial state of the field is a coherent state? Then $C_n(0)$ follow the Poissonian distribution. Unfortunately, there is no closed form for the population inversion, but I can tell you what happens. The atom still oscillates up and down between its two levels. However, the envelope of the oscillation collapses to zero when the correlation of the frequencies for the different excitations is lost. The time scale for collapse is equal to the inverse of the spread of frequencies present in the coherent state. After some time there is a revival of oscillations, because the correlation between frequencies is restored. The revival occurs as a consequence of the quantum nature of the radiation. If a continuum distribution of photons were considered the oscillations would collapse but the revival would not be observed (just like in any irreversible emission due to coupling to a continuum as in Wigner Wiesskopf treatment earlier). Collapses and revivals are our best proof that the electromagnetic field is quantized.

What about experimental confirmation of collapses and revivals? The practical relevance of the Jaynes–Cummings model may be doubtful since it describes an idealized situation of a two level system interacting with just one mode of the field. Fortunately, in a real atom the selection rules limit the allowed transitions between states so in some situations a certain state may couple to only one other. Optical pumping techniques allow such preferred states to be prepared in the laboratory and successfully used in experiments. An experimental demonstration of the model was achieved using a cooled high-Q microwave cavity (allowing us to focus on only one relevant mode of light inside) and a low density beam of Rydberg atoms. Rydberg atoms where used because they have long lived states and large dipole couplings (Rabi flopping was confirmed by the group of Haroche in Paris, Phys. Rev. Lett., **76**, 1800 (1996)).

10.3 Spontaneous Emission — At Last

So finally we are able to derive spontaneous emission from quantum mechanics. In order to do so we have had to go a long way. We had to learn about time dependent perturbation theory and how to quantize light (or any other field for that matter). Only the fully quantized model is able to do so. And coherent excitations of a monochromatic field with a two level atom is not enough either as we can guess from the semi-classical treatment. We need a bunch of modes (harmonic oscillators) all interacting incoherently with the atom via the Jaynes–Cummings interaction analyzed previously. Here is the interaction Hamiltonian representing this, so far most complicated, situation we have encountered here:

$$H_I = ie \sum_k \frac{\hbar \omega_k}{2\epsilon_0 V} d_{12} (a_k e^{-i\omega_k t}|2\rangle\langle 1| - a_k^\dagger e^{i\omega_k t}|1\rangle\langle 2|) \quad (10.35)$$

where, as before, $\mathbf{d}_{12}$ is the dipole element between the two levels 1 and 2 and all the other terms have their usual meaning as well. I have dropped the vector notation for the modes as it is no longer necessary (i.e. I am no longer using boldface) — but it should, in principle, be there. Now let's look at the matrix elements for the photon absorption and emission. Remember that we should be reading from right to left like the Arabic or Hebrew language! So, for absorption we have

$$\langle n_k - 1, 2|H_I|n_k, 1\rangle \quad (10.36)$$

$$= ie \frac{\hbar \omega_k}{2\epsilon_0 V} d_{12} e^{-i\omega_k t} \langle n_k - 1, 2|a_k|n_k, 1\rangle \quad (10.37)$$

But a is the lowering operator and hence

$$\langle n_k - 1, 2|H_I|n_k, 1\rangle = ie \frac{\hbar \omega_k}{2\epsilon_0 V} d_{12} e^{-i\omega_k t} \sqrt{n_k} \quad (10.38)$$

Similarly, for emission the element will be

$$\langle n_k + 1, 1|H_I|n_k, 2\rangle = -ie \frac{\hbar \omega_k}{2\epsilon_0 V} d_{12} e^{i\omega_k t} \sqrt{n_k + 1} \quad (10.39)$$

And here is the key result. Even when there are no photons in any of the light modes, i.e. all $n_k = 0$, the matrix element for the emission does not disappear. This feature has no (semi-)classical analogue — classically, when there is no field there is no interaction. But not so quantum mechanically. This residual interaction is responsible for the spontaneous emission. Assuming that the atom is in the state $|2\rangle$ initially, we can derive the rate of transition to the state $|1\rangle$, in complete analogy with the semi-classical case, to be

$$\frac{d}{dt}|c_1|^2 = \frac{\pi e^2 \omega_k}{\hbar \epsilon_0 V}|d_{12}|^2 \delta(\omega_{12} - \omega_k) \qquad (10.40)$$

Thus the Einstein A coefficient is obtained by summing over all possible modes with no excitations

$$A_{21} = \frac{\pi e^2}{\hbar \epsilon_0 V} \sum_k \omega_k |d_{12}|^2 \delta(\omega_{12} - \omega_k) \qquad (10.41)$$

Changing the summation to integration as before:

$$A_{21} = \frac{\pi e^2}{\hbar \epsilon_0 V} \int \frac{V \omega_k^3}{\pi^2 c^3}|d_{12}|^2 \delta(\omega_{12} - \omega_k) d\omega_k \qquad (10.42)$$

$$= \frac{e^2 \omega_{12}^3 |d_{12}|^2}{3 \hbar \epsilon_0 \pi c^3} \qquad (10.43)$$

where the factor $1/3$ comes from averaging over all possible orientations of d_{12} (three spatial directions, all equally likely). This result is identical to our previously obtained expression from the combination of the semi-classical theory and Einstein's treatment. The B coefficient is, of course, the same as in the semi-classical derivation — as we have said its value does not depend on the quantization of the light field.

10.4   The Lamb Shift

If a two level atom starts from the excited state and there are no photons around, then it will decay to the lower state and the rate is Γ. So we know the probability for the atom not to decay is $e^{\Gamma t}$. But this does not tell us about the amplitude change of the excited state. We can make a guess

$$C_b(t) = e^{-\Gamma t/2} e^{-i\omega_{ba} t} \qquad (10.44)$$

But, this is not the right form although the mod square of it reproduces the right probability. So, if we want to have the right probability we can only multiply the above expression by a unit modulus complex number, i.e. a phase. We can have the form

$$C_b(t) = e^{-\Gamma t/2} e^{-i\omega_{ba} t} e^{i\delta\omega_{ba} t} \qquad (10.45)$$

It turns out that the added correction is observable and it is manifested as an effective shift of the excited state $\omega_{ba} \to \omega_{ba} + \delta\omega_{ba}$, known as the Lamb shift (Nobel Prize, 1955). Can we calculate this quantity? The answer is yes and the shift can be calculated in a relatively simple way. Vacuum fluctuations exert a force on an electron. This changes the potential which the electron feels:

$$V(r) \to V(r + \delta r) \qquad (10.46)$$

To see the electron energy shift we need to calculate

$$V(\delta r) = V(r + \delta r) - V(r) = \nabla V \delta r + \frac{1}{2} \frac{\partial}{\partial r_i} \frac{\partial}{\partial r_j} V \delta r_i \delta r_j \quad (10.47)$$

On average, however, the first term disappears and the second term becomes

$$\langle \delta V \rangle = \frac{1}{6} \langle \nabla^2 V \rangle \langle (\delta r)^2 \rangle \qquad (10.48)$$

We now need to calculate two averages for the above formula. We have for the electron

$$m \frac{d^2 \mathbf{r}}{dt^2} = e\mathbf{E} \qquad (10.49)$$

Thus we can conclude that

$$\delta r_k \approx \frac{e}{mc^2 k^2} E_k \qquad (10.50)$$

The mean square fluctuation in position of a free electron is obtained by summing up over all the modes, so that we have

$$\langle (\delta r)^2 \rangle = \frac{2}{\pi} \frac{e^2}{\hbar c} \left(\frac{\hbar}{mc} \right)^2 \int_{k_0}^{k} \frac{dk}{k} \qquad (10.51)$$

We also take the potential to be

$$V = -\frac{e^2}{4\pi\epsilon_0 r} \qquad (10.52)$$

We now need to calculate

$$\langle \nabla^2 V \rangle = -\frac{e^2}{4\pi\epsilon_0} \left\langle \nabla^2 \frac{1}{r} \right\rangle \qquad (10.53)$$

I now use the fact that[1]

$$\nabla^2 \frac{1}{r} = -4\pi\delta(r) \tag{10.54}$$

so that

$$\langle \nabla^2 V \rangle = -\frac{e^2}{4\pi\epsilon_0} \int dr \psi^* \nabla^2 \frac{1}{r} \psi \tag{10.55}$$

$$= \frac{e^2}{\epsilon_0} |\psi(0)|^2 \tag{10.56}$$

This contribution vanishes if the wave function is not finite at the origin. For the S state this is true, but it disappears for the P state. For the $2S$ state it is

$$\langle \nabla^2 V \rangle_{2S} = \frac{e^2}{8\pi\epsilon_0 a_0^3} \tag{10.57}$$

This can be calculated to yield the difference between the s and p state

$$\Delta E = \langle \delta V \rangle = \frac{\hbar e^4}{12\pi^2 \epsilon_0^2 m^2 c^3} |\psi_s(0)|^2 \ln \frac{mc^2}{\hbar\omega_0} \tag{10.58}$$

For the $2S$ state of hydrogen, $\Delta E/\hbar = 1040 MHz$ and was confirmed by Lamb and Rutherford in 1947.

10.5 Parametric Down Conversion

Now we will revise some non-linear processes in the light of quantizing electromagnetic field. If you remember, one non-linear process we covered was second harmonic generation. From the quantum perspective this involves two photons of energy ω combining to produce a single photon of energy 2ω. Down conversion is, loosely speaking, the opposite of this process. Here we have a single photon of frequency ω_0 incident on a crystal with second order non-linearity $\chi^{(2)}$, which generates two photons with frequencies ω_1 and ω_2, such that

$$\omega_0 = \omega_1 + \omega_2 \tag{10.59}$$

In the traditional jargon, ω_0 is known as the pump frequency and the resulting ω_1 and ω_2 are known as the signal and idler. So, how

---

[1]This can be derived from the first Maxwell equation $\nabla E = \rho/\epsilon$. We can express the field E via the vector potential, so that $-\nabla\phi = E$, in which case we have that $\nabla^2\phi = \rho/\epsilon$. Now, for the point charge we have $\rho = e\delta(r)$, and since $\phi = e/4\pi\epsilon r$, we have that $\nabla^2(1/r) = -4\pi\delta(r)$.

do we describe this process quantum mechanically? We need to know the Hamiltonian, which is

$$H = \sum_{i=0}^{2} \hbar\omega_i \left(n_i + \frac{1}{2} \right) + \hbar g(a_1^\dagger a_2^\dagger a_0 + a_1 a_2 a_0^\dagger) \qquad (10.60)$$

Let's see why this is the correct Hamiltonian. The initial state before the crystal is $|\Psi(t=0)\rangle = |0_{\omega_1}, 0_{\omega_2}, 1_{\omega_0}\rangle$ indicating one photon at ω_0 and none at ω_1 and ω_2. By solving the Schrödinger equation we obtain (this is the same as the Jaynes–Cummings model, but with three, instead of two systems) the following:

$$\cos gt|0_{\omega_1}, 0_{\omega_2}, 1_{\omega_0}\rangle + i \sin gt|1_{\omega_1}, 1_{\omega_2}, 0_{\omega_0}\rangle \qquad (10.61)$$

Therefore, after $gt = \pi/2$, we have the state

$$|1_{\omega_1}, 1_{\omega_2}, 0_{\omega_0}\rangle \qquad (10.62)$$

which is the state with two generated photons.

10.6 Quantum Measurement: A Brief Discussion

I'd like to close this section with a bit more philosophical topic. This topic is very appropriate at this stage as we realize that everything around us — both matter and fields (radiation) — has quantum features which we can, in fact, confirm experimentally to a high degree of reliability. So the question is: how come that the world we live in "looks so classical"? Where have all the superpositions (on the macroscopic level) disappeared? Or, have they? There is much discussion about these and related topics and we don't have enough space in a book of this type to talk about all the different views on this problem. It is related to the issue of quantum measurement which I would like to say a few words about now.[2]

Let's look at a superposition of two different quantum states:

$$|\psi\rangle = |\psi_1\rangle + |\psi_2\rangle \qquad (10.63)$$

---

[2]There is a huge literature written on this topic. I strongly recommend the collection of papers by Wheeler and Zurek [Wheeler and Zurek (1992)]. Some people would even say that this is one of the key open problems at the foundations of quantum mechanics and that it gives rise to the irreversibility in the second law, or the arrow of time or the quantization of gravity or what not. My view is that, once you look closer at it the problem completely disappears and we don't really need to worry about it all that much. I won't, however, be trying to convince you of my point of view here.

What do we mean by the concept "coherence" between the states $|\psi_1\rangle$ and $|\psi_2\rangle$? What I mean by this question is to ask how a superposition is different from just having either one or the other state (in the classical probabilistic sense). This is best seen from comparing the two alternatives in an experiment. Suppose that we want to compute the average of some operator $\hat{O}$. According to the rules of quantum mechanics, it is

$$\langle\hat{O}\rangle = \langle\psi_1|\hat{O}|\psi_1\rangle + \langle\psi_2|\hat{O}|\psi_2\rangle + 2|\langle\psi_1|\hat{O}|\psi_2\rangle|\cos\phi \qquad (10.64)$$

where ϕ is the phase of the complex number

$$\langle\psi_1|\hat{O}|\psi_2\rangle = |\langle\psi_1|\hat{O}|\psi_2\rangle|e^{i\phi} \qquad (10.65)$$

The existence of this term is called coherence. It leads to all the weird quantum effects, some of which we have encountered in this book thus far. When this term disappears we have a classical mixture of states as a result. The mixture of states and the coherent superposition are really two different physical states and this is where the difference between the classical and quantum physics lies. Once you understand this, the rest is just the same game, but played with more than one system.

This disappearance of the coherent term is also what happens during a quantum measurement,[3] which is why the two issues of "classicality" and "quantum measurement" are related. Now we present a fully quantum model for a measurement. This was first done by von Neumann in his seminal book entitled *The Mathematical Foundations of Quantum Mechanics*. It was later used by Everett [Everett (1973)] to argue for the "many worlds picture" of quantum mechanics. We treat the measurement apparatus as another quantum system and represent its initial state by $|m\rangle$. This is intellectually pleasing if we believe that quantum mechanics is a universally valid theory and there is no reason not to. The purpose of the measurement is to "entangle" the state of the apparatus with that of the system like so:

$$(|\psi_1\rangle + |\psi_2\rangle)|m\rangle \rightarrow |\psi_1\rangle|m_1\rangle + |\psi_2\rangle|m_2\rangle \qquad (10.66)$$

The average is now given by

$$\langle\hat{O}\rangle = \langle\psi_1|\hat{O}|\psi_1\rangle + \langle\psi_2|\hat{O}|\psi_2\rangle + 2\Re(\langle\psi_1|\hat{O}|\psi_2\rangle\langle m_1|m_2\rangle) \qquad (10.67)$$

---

[3]There is nothing more I would add to the issue of quantum measurement, but like I said, the jury is still out on this question and there is a great debate going on in some parts of the quantum community regarding the exact meaning of measurement.

Note that if the states of the apparatus are orthogonal (what is called a perfect measurement), the coherence disappears and the average value of $\hat{O}$ is just the average of its values in the individual states ψ_1 and ψ_2.

The state $|\psi_1\rangle|m_1\rangle + |\psi_2\rangle|m_2\rangle$ has been crucial in the development of quantum mechanics and a source of a great deal of discussions, paradoxes and controversy. It is known as an *entangled* state. The state is called entangled because it cannot be written as a product of the individual states of the two systems separately. So, loosely speaking, the systems lose their identity when they are entangled. An example of an entangled state is the atom light state in the Jaynes–Cummings model, e.g. $|e,0\rangle + |g,1\rangle$. Entanglement is a source of many (apparent) paradoxes in quantum mechanics and understanding it precisely still eludes current researchers in the field [Vedral (2002)].

Chapter 11

Some Recent Applications of Quantum Optics

There are many applications of lasers and the formalism we have learnt about here. I have chosen two recent applications which are very relevant and exciting (one of them led to two Nobel Prizes in last five years). In the last part of the book I will speak about the most exciting of all applications: quantum information processing. (You can read more on quantum information processing in my review [Vedral (2002)].)

11.1 Laser Cooling

We have seen that a resonant interaction of light with an atom leads to absorption during which momentum is transferred to the atom. If the atom emits spontaneously, this then on average leads to no recoil as spontaneous emission is isotropic. Suppose that the beam of light is shone on an atom moving in a certain direction. When a photon is absorbed, the atom slows down and the velocity reduction $\Delta \mathbf{v}$ is inferred from

$$M\Delta \mathbf{v} = \hbar \mathbf{k} \tag{11.1}$$

where M is the atom's mass and $\mathbf{k}$ the photon's wave vector. Let's estimate how many photon absorptions we need to bring an atom to a standstill. The root mean square velocity is given by

$$v = \left(\frac{3k_B T}{M} \right)^{1/2} \tag{11.2}$$

At $T = 600K$, $v \approx 800m/s$. It takes about

$$N = v/\Delta v \approx 13000 \tag{11.3}$$

cycles of emission and absorption to stop this atom. However, the atom will never be stopped in reality. This is because the radiation will provide random kicks for the atom to undergo a form of random motion.[1] Let's discuss this a bit more. What is the distance travelled by an atom in time t if it "walks" randomly? Let's suppose that the atom has already made n steps and that the distance it moved is $\mathbf{r}_n$. Suppose further that the move in each step is $\delta\mathbf{r}$, then we can write the following recursive relationship:

$$\mathbf{r}_{n+1} = \mathbf{r}_n + \delta\mathbf{r} \tag{11.4}$$

This is all obvious, but now comes the smart bit. First we square this equation (you'll see in a minute why):

$$\mathbf{r}_{n+1}^2 = \mathbf{r}_n^2 + 2\mathbf{r}_n\delta\mathbf{r} + \delta^2\mathbf{r} \tag{11.5}$$

We cannot know what the distance will actually be, but on average we can say what we *expect* it to be, hence,

$$\langle\mathbf{r}_{n+1}^2\rangle = \langle\mathbf{r}_n^2\rangle + 2\langle\mathbf{r}_n\delta\mathbf{r}\rangle + \langle\delta^2\mathbf{r}\rangle \tag{11.6}$$

However, since the total length of motion and each individual step are independent, we have that $\langle\mathbf{r}_n\delta\mathbf{r}\rangle = \langle\mathbf{r}_n\rangle\langle\delta\mathbf{r}\rangle$. Also, since we have a truly random motion this by definition implies that $\langle\delta\mathbf{r}\rangle = 0$. Thus,

$$\langle\mathbf{r}_{n+1}^2\rangle = \langle\mathbf{r}_n^2\rangle + \langle\delta^2\mathbf{r}\rangle \tag{11.7}$$

Iterating this formula gives

$$\langle\mathbf{r}_n^2\rangle = n\langle\delta^2\mathbf{r}\rangle \tag{11.8}$$

Now, $n = t/\delta t$ and $\langle v^2\rangle = \langle\delta^2\mathbf{r}\rangle/\delta^2 t$. So we obtain

$$\langle\mathbf{r}^2(t)\rangle = \beta t \tag{11.9}$$

where β is the diffusion coefficient

$$\beta = \langle v^2\rangle\delta t \tag{11.10}$$

From the equipartition theorem we know that $\langle v^2\rangle = 3kT/m$, so that

$$\langle\mathbf{r}^2(t)\rangle = \frac{3kT}{m}t\delta t \tag{11.11}$$

Therefore, by measuring the distance travelled by an atom in time t, we can determine the lowest temperature of cooling. Let's push

---

[1]This is like Brownian motion, but in the case of Brownian motion atoms in the surrounding gas provide kicks for a speck of dust to move around.

this a bit further. First of all, the average time between the kicks is equal to the life-time of the excited state, i.e.

$$\delta t = \frac{2\pi}{\Gamma} \tag{11.12}$$

On the other hand, the momentum of the atom can be written in two different ways as

$$\frac{m\sqrt{\langle r^2 \rangle}}{t} = p = \hbar k = \hbar \frac{2\pi}{\sqrt{\langle r^2 \rangle}} \tag{11.13}$$

Therefore, this implies that the temperature of this randomly moving atom is

$$T = \frac{\hbar \Gamma}{3k} \tag{11.14}$$

This is the so-called Doppler limit for cooling. This leads to temperatures of about $T_{\min} = 1mK$ in practice. We need to go 10^6 times below this to achieve condensation.

11.1.1 *Bose–Einstein condensation*

Bose–Einstein condensation is a process where quantum systems obeying Bose statistics, say atoms, reach a state where their identity becomes completely obscured and they all start to behave like a single, albeit large, quantum system. This impressive phenomenon, predicted by Bose and Einstein in 1918, was observed only in 1995 (in gases of atoms) and was awarded a Nobel Prize in physics a few years ago.[2]

Experimentally, an atom condensate is very difficult to obtain. Observation of quantum statistics occurs only at very low temperatures (at high temperatures all particles follow the Maxwell–Boltzmann distribution). Achieving this temperature involves combining two separate experimental techniques. First is laser cooling described above. But this can only reach the milliKelvin regime, i.e. it reduces the speed of atoms from $800m/s$ to about $10m/s$. To obtain a clear condensate we need to go to 10^6 times lower! Let's perform a "back-of-an-envelope" calculation to estimate this temperature. According to de Broglie's hypothesis the relationship between atom's velocity v and wavelength λ is

$$v = \frac{\hbar}{m\lambda} \tag{11.15}$$

---

[2]For a very recent introduction to Bose–Einstein condensation see [Pitaevskii (1993)].

If we want to condense N atoms each of size a_0, then the radius R of the condensate is roughly

$$R = a_0 \sqrt{N} \qquad (11.16)$$

Now we want the temperature at which

$$R \approx \lambda \qquad (11.17)$$

which implies that

$$v = \frac{\hbar}{ma_0\sqrt{N}} \qquad (11.18)$$

According to equipartition of energy,

$$\frac{3}{2}kT = \frac{mv^2}{2} \qquad (11.19)$$

Thus,

$$T = \frac{1}{3k}\frac{\hbar^2}{ma_0^2 N} \qquad (11.20)$$

For $N = 1000$ atoms this comes out to be $T = 10^{-11}K$. This is an overestimate, i.e. the atoms don't need to be quite that cold, but should be in the regime of nanoKelvin. For this we need to trap laser cooled atoms and then cool them by evaporation. The technique which we use to get extremely cold atoms is called rf-induced evaporation. It is very similar to the way evaporation works in a cup of hot coffee. A cup of coffee is made up of many molecules flying around and bumping into each other. The temperature of the coffee is just a measure of the average energy that these quickly moving molecules have. From time to time two molecules will collide in such a way that one of the two ends up with most of the energy, sometimes even gaining enough energy to fly out of the cup. Since these molecules are going relatively fast compared to the rest of the molecules, they take with them more than their fair share of energy, and the molecules which are left behind have less energy on average than they did before the fast molecules shot out. For every molecule that is kicked out, the temperature of the coffee decreases a tiny amount.

I will now use the same statistical reasoning we used for photons to explain the Bose–Einstein phenomenon. The fact that the Bose particles all behave like photons (even though they could be atoms, for example) has been one of the great unifications of quantum

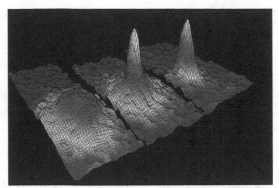

This figure is reproduced from the JILA website
http://www.colorado.edu/physics/2000/bec/three_peaks.html
with the full permission of the authors.

Fig. 11.1 Velocity distribution of atoms as they approach the Bose–Einstein condensate. The transition to the lowest state of motion in the trap is beautifully seen in this plot which resulted in the Nobel Prize for the authors.

mechanics. Suppose that atoms are trapped in a "square box" of side L. The total number of particles is

$$N = \sum_{\alpha} n_{\alpha} \qquad (11.21)$$

where

$$n_{\alpha} = \frac{1}{e^{\beta E_{\alpha} - \beta \mu} - 1} \qquad (11.22)$$

The average number of particles in the single particle state α must be positive, so that $\mu \leq E_{\alpha}$. Now we have that the density is (converting the above sum into an integral as usual)

$$n = \frac{N}{V} = \int \frac{dp}{h^3} \frac{1}{e^{\beta E_p - \beta \mu} - 1} \qquad (11.23)$$

This integral can be written in the dimensionless form as

$$n = 4\pi \left(\frac{mkT}{h^2} \right)^{3/2} \int dx x^2 \frac{1}{e^{x^2/2 - \beta \mu} - 1} \qquad (11.24)$$

This expression reaches its maximum value when $\mu = 0$ when we have

$$n_{\max} = 4\pi \left(\frac{mkT}{h^2} \right)^{3/2} \int dx x^2 \frac{1}{e^{x^2/2} - 1} \qquad (11.25)$$

But how can this be? These particles are bosons and we are saying that there is a limit to how many of them we can have in a box. What is preventing us from adding another boson? The answer is

— nothing. We have, in fact, made a mistake in our calculation and did not take into account the lowest energy state for which

$$n_0 = \frac{1}{e^{-\beta\mu} - 1} \qquad (11.26)$$

When the temperature decreases, this can be made arbitrarily large. So we can have as many atoms as we like in the ground state. In fact, the total number of atoms is

$$N = n_0 + V n_{\max} \qquad (11.27)$$

For low temperature $n_{\max} \ll n_0$ and all the particles go to the ground state — which is, of course, Bose–Einstein condensation (see Figure 11.1 for experimental results of the first atomic condensation).

Final Thought: At the end, I will briefly tell you about a fascinating discovery of Dirac's which signified the introduction of quantum field theory. (For a modern introduction to quantum field theory see [Weinberg (1997)].) It is the fact that N quantum systems that are bosons (it could be N atoms in a trap forming a Bose–Einstein condensate) behave like N harmonic oscillators. So, not only is light a harmonic oscillator, but also so is any collection of bosons (e.g. atoms with integral spin). The treatment that exposes this formal analogy is called the *second quantization* for historical reasons. This is the point where quantum optics stops and quantum field theory begins. Very briefly speaking, any continuous function of space and time can be thought of as a field. The quantum wave function, $\Psi(x,t)$, is one such function and we can also think of it as a field. As with any field, we can write it in terms of creation and annihilation operators, so that we would have $\Psi(x,t) = \sum_k e^{i(kx-\omega t)}a(k) + e^{-i(kx-\omega t)}a^\dagger(k)$ (just like the electric field is a sum of creation and annihilation operators). The point is that these creation and annihilation operators now describe creating an atom and annihilating an atom with momentum k. The wave function is therefore also now an operator and it describes annihilating an atom at position x and time t ($\Psi(x,t)^\dagger$ on the other hand describes creating an atom at x and t).

11.2 Quantum Information Processing

Quantum physics not only provides the most complete description of physical phenomena known to science, it also provides a new philosophical framework for our understanding of Nature. It enables us to accurately model microscopic systems such as quarks and atoms to large cosmic objects such as black holes. Information theory, on the other hand, teaches us about our physical ability to store and process information. Without a formalized information theory many of the recent developments in telecommunications, computer science and engineering would simply not have been possible. Although quantum physics and information theory initially developed separately, their recent integration is seen as yet another important step towards understanding the fundamental properties and limitations of Nature. And the exciting thing is that quantum mechanics allows us to perform some information processing tasks that cannot be done with classical physics (by classical I mean anything that is not quantum, and this includes general relativity). Here is one striking example — quantum teleportation.

11.2.1 *Quantum teleportation*

Let us begin by describing quantum teleportation in the form originally proposed by Bennett *et. al* (Phys. Rev. Lett. **70**, 1895 (1993)). Suppose that Alice and Bob, who are distant from each other, wish to implement a teleportation procedure. Initially they need to share a maximally entangled pair of qubits. This means that if Alice and Bob both have one qubit each, then the joint state may, for example, be the Einstein–Rosen–Podolsky pair [Einstein, Podolsky and Rosen (1935)]:

$$|\Psi_{AB}\rangle = (|0_A\rangle|0_B\rangle + |1_A\rangle|1_B\rangle)/\sqrt{2} \qquad (11.28)$$

where the first ket (with subscript A) belongs to Alice and second (with subscript B) to Bob. Note that this state is maximally entangled and is different from a statistical mixture $(|00\rangle\langle00|+|11\rangle\langle11|)/2$ which is the most correlated state allowed by classical physics.

Now suppose that Alice receives a qubit in an unknown state $|\Phi\rangle = a|0\rangle + b|1\rangle$ (the state is unknown if she doesn't know the values of a and b) and she wants to teleport it to Bob. The state has to be unknown to her because otherwise she can just phone Bob up and tell him all the details of the state, and he can then recreate it on a particle that he possesses. Given that Alice does not know the state, she cannot measure it to obtain all the necessary information to specify it. If she could, this would lead to a violation of uncertainty principle. Therefore she has to resort to using the state $|\Psi_{AB}\rangle$ that she shares with Bob to transfer her state to him without actually learning this state. This procedure is what we mean by quantum teleportation.

I first write out the total state of all three qubits:

$$|\Phi_{AB}\rangle := |\Phi\rangle|\Psi_{AB}\rangle = (a|0\rangle + b|1\rangle)(|00\rangle + |11\rangle)/\sqrt{2} \quad (11.29)$$

However, the above state can be conveniently written in a different basis:

$$\begin{aligned}
|\Phi_{AB}\rangle &= (a|000\rangle + a|011\rangle + b|100\rangle + b|111\rangle)/\sqrt{2} \\
&= \frac{1}{2}[|\Phi^+\rangle(a|0\rangle + b|1\rangle) + |\Phi^-\rangle(a|0\rangle - b|1\rangle) \\
&\quad + |\Psi^+\rangle(a|1\rangle + b|0\rangle) + |\Psi^-\rangle(a|1\rangle - b|0\rangle)]
\end{aligned}$$

where

$$|\Phi^+\rangle = (|00\rangle + |11\rangle)/\sqrt{2} \qquad\qquad (11.30)$$

$$|\Phi^-\rangle = (|00\rangle - |11\rangle)/\sqrt{2} \qquad\qquad (11.31)$$

$$|\Psi^+\rangle = (|01\rangle + |10\rangle)/\sqrt{2} \qquad\qquad (11.32)$$

$$|\Psi^-\rangle = (|01\rangle - |10\rangle)/\sqrt{2} \qquad\qquad (11.33)$$

form an orthonormal basis of Alice's two qubits (remember that the first two qubits belong to Alice and the last qubit belongs to Bob). The above basis is frequently called the Bell basis. This is a very useful way of writing the state of Alice's two qubits and Bob's single qubit because it displays a high degree of correlation between Alice's and Bob's parts: for every state of Alice's two qubits (i.e. $|\Phi^+\rangle, |\Phi^-\rangle, |\Psi^+\rangle, |\Psi^-\rangle$) there is a corresponding state of Bob's qubit. In addition, the state of Bob's qubit in all four cases looks very much like the original qubit that Alice has to teleport to Bob. It is now straightforward to see how to proceed with the teleportation protocol:

(1) Upon receiving the unknown qubit in state $|\Phi\rangle$, Alice performs projective measurements on her two qubits in the Bell basis. This means that she will obtain one of the four states in the Bell basis randomly and with equal probability.

(2) Suppose Alice obtains the state $|\Psi^+\rangle$. Then the state of all three qubits (Alice⊗Bob) collapses to the following state:

$$|\Psi^+\rangle(a|1\rangle + b|0\rangle) \qquad\qquad (11.34)$$

(the last qubit belongs to Bob as usual). Alice now has to communicate the result of her measurement to Bob (over the phone, for example). The point of this communication is to inform Bob how the state of his qubit now differs from the state of the qubit Alice was holding before the Bell measurement (just a measurement in the Bell basis).

(3) Now Bob has to apply a unitary transformation on his qubit which simulates a logical NOT operation: $|0\rangle \rightarrow |1\rangle$ and $|1\rangle \rightarrow |0\rangle$. He thereby transforms the state of his qubit into the state $a|0\rangle + b|1\rangle$, which is precisely the state that Alice had to teleport to him initially. This completes the protocol. It is easy to see that if Alice obtained some other Bell state, then Bob would have to apply some other simple operation to complete the teleportation. They can be represented by the Pauli spin matrices.

An important fact to observe in the above protocol is that all the operations (Alice's measurements and Bob's unitary transformations) are *local* in nature. This means that there is never any need to perform a (global) transformation or measurement on all three qubits simultaneously, which is what allows us to call the above protocol a genuine teleportation. It is also important that the operations that Bob performs are independent of the state that Alice tries to teleport to him. Note also that the classical communication from Alice to Bob in step 2 above is crucial because otherwise the protocol would be impossible to execute (there is a deeper reason for this: if we could perform teleportation without classical communication then Alice could send messages to Bob faster than the speed of light).

It is important to observe that the fact that the initial state to be teleported is destroyed immediately after Alice's measurement, i.e it becomes maximally mixed of the form $(|0\rangle\langle 0| + |1\rangle\langle 1|)/2$. This has to happen since otherwise Alice and Bob would end up with two qubits in the same state. So, effectively, they would clone an unknown quantum state, which is impossible by the laws of quantum mechanics. This is the no-cloning theorem of Wootters and Zurek (published in 1982), which is a simple consequence of *linearity* of quantum dynamical laws. We also see that at the end of the protocol the quantum entanglement of $|\Psi_{AB}\rangle$ is completely destroyed. Does this have to be the case in general or might we save that state at the end (by perhaps performing a different teleportation protocol)? The answer is that the entanglement must be destroyed and the reason is that if this was not the case, then entanglement could increase under local operations and classical correlations, which is taken to be prohibited by definition (recall that two systems are entangled if non-classical correlations exist between them).

Teleportation has been experimentally performed in many different set-ups (see, for example, Bouwmeester *et. al, Experimen-*

tal quantum teleportation, Nature **390**, 575 (1997)). Currently, it should be emphasized, we can only teleport individual quantum systems, and the way to larger objects (flies, for example?) seems to be long and hard. For more on this and quantum computing in general I warmly recommend Deutsch's book — *The Fabric of Reality* [Deutsch (1997)].

Chapter 12

Closing Lines

My intention in this book has been to explain both the fundamental side of the behavior of light, as well as introduce some exciting recent developments. The subject of quantum optics is very much alive and it is a rapidly developing field. It currently involves activities such as testing the foundations of quantum mechanics, i.e. the limitations — if any — of the superposition principle, as well as the activity of trying to find an appropriate medium to implement quantum computation.

We shouldn't, of course, just emphasize the "testing and building" side of quantum optics. Knowledge should also be gained for its own sake. With every new physical theory comes a new understanding of the world we live in. Through Newtonian physics we understood the Universe as a clockwork mechanism. With the subsequent development of thermodynamics the Universe became a big Carnot engine, slowly evolving towards its final equilibrium state after which no useful work could be obtained — the heat death. Presently, we see the universe as an information processing machine, storing and manipulating information through quantum systems such as photons and atoms — a quantum computer, in other words.

If there is a single moral to be drawn from this book on quantum optics it is that, as we dig deeper into the fundamental laws on physics, we also push back the boundaries of our understanding. It will not be surprising at all if all the results presented in this book are superseded by a higher level generalization of which they become an approximation, in the same way that today classical optics approximates the more accurate quantum optics.

Chapter 13

Problems and Solutions

This section contains five sets of problems together with detailed solutions. The problems have mainly been taken from exam papers that I set at Imperial College London. The first two problem sets can be attempted with the knowledge of the first seven chapters only. For the third problem set the knowledge of Chapters 8 and 9 will be required, while sets four and five require knowledge of Chapters 10 and 11.

13.1 Problem and Solutions 1

13.1.1 *Problem set 1*

(1) This question asks you to estimate the temperature of the Earth by assuming that it is a blackbody. (The final result is pretty close to the actual average temperature of about $300K$.) Assume that the solar radiation flux is spherically symmetric and that its value at Earth is $F = 1.37kWm^{-2}$. Assuming also that 30 percent of the radiation is scattered by the atmosphere back into space (i.e. it doesn't get to Earth's surface), argue that the total amount of heat absorbed by the Earth is

$$(1 - 0.3)\pi a^2 F \tag{13.1}$$

where a is Earth's radius. Since our planet is in thermal equilibrium, its (average) temperature is pretty much constant (forget the nonsense about "global warming"), and so all this heat is re-emitted (otherwise the temperature would increase). Use the Stefan–Boltzmann law to argue that the emitted heat is

$$4\pi a^2 \sigma T^4 \tag{13.2}$$

where σ is Stefan's constant. In equilibrium, the emitted and absorbed heats have to be equal. Use this fact to calculate the

resulting temperature of our planet. The value that you will
obtain is almost the same as that of troposphere. What do
you think is the main reason why Earth's surface is some $40K$
hotter? (Notice that the value of the radius of our planet is
not relevant in this question!)

(2) Derive equations (3.34) and (4.15) from the text.

(3) (i) Explain how the Einstein rate coefficients describe absorption, stimulated emission and spontaneous emission in a two
state system. (ii) Using Einstein's rates, describe how the populations in steady state in a two level system driven by radiation of mean energy density $\langle W \rangle$ saturate. (iii) A $20mW$ laser
beam with a bandwidth of less than $100kHz$ is focused to a
spot of $100\mu m$ in diameter; calculate the intensity of the laser
and the electric field of the laser at the spot.

(4) Derive the condition for lasing by answering the following ques-
tions.

   (i) Write down the appropriate rate equations for a two level
   system (levels labelled as 1 and 2) taking into account
   both the blackbody radiation and an external source.

   (ii) Solve this equation to derive the population as a function
   of time in both levels assuming that all the population is
   initially in the ground state 1.

   (iii) By assuming the steady state and absence of the external
   radiation source derive the relationship between the A
   and B coefficients in terms of the transition frequency,
   ω_{12} between the levels 1 and 2. You may assume that
   Planck's blackbody radiation density is given by

   $$W_T = \frac{\hbar\omega_{12}^3}{\pi^2 c^2} \frac{1}{e^{\hbar\omega_{12}/kT} - 1} \qquad (13.3)$$

   where all the symbols have their usual meaning.

   (iv) What is the condition for lasing?

   Show that a two level system cannot achieve lasing.

(5) Suppose that in the Einstein two level model, the two levels
a and b are degenerate (this goes beyond Einstein's original
treatment) with the respective degeneracies g_a and g_b. What
is the relationship between the stimulated emission and absorption rates now? What about the relationship between the
A and the B coefficient?

(6) What are the main features of the blackbody spectrum?
The probability of there being n photons in the mode with
frequency ω inside a blackbody at temperature T is given by

the Boltzmann formula. The density of states in a photon gas is given by

$$\rho(\omega) = \frac{\omega^2}{\pi^2 c^3} \tag{13.4}$$

Show that the Planck blackbody formula is given by the following expression

$$W_T = \frac{\hbar \omega^3}{\pi^2 c^3} \frac{1}{e^{\hbar \omega / kT} - 1} \tag{13.5}$$

Identify all the terms in the above formula.

(i) Calculate the standard deviation in the photon number in terms of the average number of photons $\langle n \rangle$.

(ii) Which term in the expression of the standard deviation carries the signature of particle behavior and which reflects wave behavior and why? (A qualitative answer is sufficient here.)

You may use the following expression:

$$\frac{\sum_{n=0}^{\infty} n e^{-n\hbar\omega/kT}}{\sum_{m=0}^{\infty} e^{-m\hbar\omega/kT}} = \frac{1}{e^{\hbar\omega/kT} - 1} \tag{13.6}$$

13.1.2 *Solutions 1*

(1) The equilibrium condition dictates that

$$\pi a^2 (1 - 0.3) F = 4\pi a^2 \sigma T^4 \tag{13.7}$$

from which it follows that

$$T = \left(\frac{0.7F}{4\sigma}\right)^{1/4} = 256K \tag{13.8}$$

This is lower than the actual average due to the atmosphere (the so-called greenhouse effect).

(2) The dispersion in Planck's distribution can be found by differentiating two different expressions for the average number of photons

$$\frac{1}{e^{\beta} - 1} = \sum_n n e^{-n\beta} (1 - e^{-\beta}) \tag{13.9}$$

by β. We obtain

$$-\frac{e^{\beta}}{(e^{\beta}-1)^2} = -\sum_n n^2 e^{-n\beta}(1-e^{-\beta}) \qquad (13.10)$$

$$+ \sum_n ne^{-n\beta}e^{-\beta} \qquad (13.11)$$

which can be rearranged to get

$$\langle n^2 \rangle = \langle n \rangle \frac{e^{\beta}}{e^{\beta}-1} \qquad (13.12)$$

$$- \langle n \rangle + \sum_n ne^{-n\beta} \qquad (13.13)$$

The first two terms sum up to yield $\langle n \rangle (1/e^{\beta} - 1) = \langle n \rangle^2$. The last terms is equal to $1/(e^{\beta}-1) \times 1/(1-e^{-\beta}) = e^{\beta}/(e^{\beta}-1)^2 = \langle n \rangle + \langle n \rangle^2$, which altogether leads to

$$\langle n^2 \rangle = \langle n \rangle + 2\langle n \rangle^2 \qquad (13.14)$$

Therefore,

$$\sigma^2 = \langle n^2 \rangle - \langle n \rangle^2 = \langle n \rangle (1 + \langle n \rangle) \qquad (13.15)$$

The evolution of population 2 can easily be derived from the rate equation. From the notes we have that

$$-\frac{dN_2}{dt} = N_2 A + (N_2 - N_1)B\langle W \rangle \qquad (13.16)$$

Now we'd like this to be a function of N_2 only, so we substitute $N_1 = N - N_2$. This leads to

$$-\frac{dN_2}{dt} = N_2 A + (2N_2 - N)B\langle W \rangle \qquad (13.17)$$

which is a simple equation that can be solved by a trial solution:

$$N_2 = \alpha e^{-\beta t} + \gamma \qquad (13.18)$$

where the coefficients α, β, γ can be determined by substituting this into the differential equation itself and then requiring that at $t = 0$ we have that $N_2 = 0$. This leads to the solution

$$N_2 = \frac{NB\langle W \rangle}{A + 2B\langle W \rangle}(1 - e^{-(A+2B\langle W \rangle)t}) \qquad (13.19)$$

as in the notes.

(3) (i) This can all be found in this book in the section on Einstein's rate equations; (ii) This is also straight from this book and I will not repeat it here; (iii) Power of the laser is given, and the power can be expressed via intensity, I, as

$$P = IA\Delta\omega \tag{13.20}$$

where $A = \pi(d/2)^2$ is the area of the spot. This formula is just the definition of power as "energy divided by time" ($\Delta\omega$ is there to take into account all the modes from the band). So the intensity (per mode) can be calculated to be

$$I = \frac{P}{A\Delta\omega} \approx 2Wm^{-2} \tag{13.21}$$

The (total) intensity is, on the other hand, equal to the speed of light times energy times density of modes, so that

$$I\Delta\omega = \frac{c\epsilon_0 E^2}{2} \tag{13.22}$$

where $\epsilon_0 E^2/2$ is the energy density. The electric field can be found to be

$$E^2 = \frac{2I\Delta\omega}{c\epsilon_0} \tag{13.23}$$

Thus, the magnitude of the field strength is $E \approx 0.3 \times 10^5 Vm^{-1}$.

(4) Einstein assumed that there are three processes involved in a two level atom interacting with light, spontaneous emission and stimulated emission and absorption. Then he assumed that the rate of change of populations is

$$\frac{dN_1}{dt} = -\frac{dN_2}{st} = B\rho(N_2 - N_1) + AN_2 \tag{13.24}$$

He showed that the Rayleigh–Jeans law implies that the rates of stimulated emission and absorption must be the same. If there was no stimulated emission then the right blackbody spectrum cannot be predicted. The main shortcoming is that the analysis is heuristic and emission and absorption rates (A and B coefficients) cannot be calculated from first principles.

(i) The equations for the two levels are

$$\frac{dN_1}{dt} = -\frac{dN_2}{st} = BW(N_2 - N_1) + AN_2 \tag{13.25}$$

where $W = W_T + W_S$.

(ii) Let's write the equation for N_1 in terms of N_1 and $N = N_1 + N_2$. We obtain

$$\dot{N}_1 = (N - N_1)A + (N - 2N_1)BW \qquad (13.26)$$

The general solution of this equation is of the form

$$N_1 = \alpha e^{-\beta t} + \gamma \qquad (13.27)$$

We plug this back into the rate equation to obtain

$$-\alpha\beta e^{-\beta t} + (A+2BW)\gamma + (A+2BW)\alpha e^{-\beta t} - (A+BW) = 0 \qquad (13.28)$$

Therefore,

$$\gamma = \frac{(A+BW)N}{A+2BW)} \qquad (13.29)$$

$$\beta = A + 2BW \qquad (13.30)$$

The constant α can be determined from the initial condition $N_1(0) = N$. Thus,

$$\alpha = \frac{BWN}{A+2BW} \qquad (13.31)$$

The final solution is

$$N_1(t) = \frac{BWN}{A+2BW}e^{-(A+2BW)t} + \frac{(A+BW)N}{A+2BW} \qquad (13.32)$$

$$= \frac{BWN}{A+2BW}(1 + e^{-(A+2BW)t}) + \frac{AN}{A+2BW} \qquad (13.33)$$

Also, the population for N_2 is immediately given as $N_2(t) = N - N_1(t)$.

(iii) Steady state implies that there is no change in populations. Then we can deduce from the rate equation

$$\frac{BW_T}{A} = \frac{N_2}{N_1 - N_2} = \frac{1}{N_1/N_2 - 1} \qquad (13.34)$$

We know that in equilibrium

$$\frac{N_1}{N_2} = e^{\frac{\hbar\omega_{12}}{kT}} \qquad (13.35)$$

therefore

$$W_T = \frac{A}{B(e^{\frac{\hbar\omega_{12}}{kT}} - 1)} \qquad (13.36)$$

Comparing this to Planck's law, we see that

$$A = \frac{\hbar\omega^3}{\pi^2 c^2}B \tag{13.37}$$

(iv) The condition for lasing is the so-called population inversion, i.e. $N_2 > N_1$.

It is clear that this cannot be achieved with a two level system as the maximum value that can be reached (when $W \to \infty$) is $N/2$.

(5) The condition is that $g_a B_{ba} = g_b B_{ab}$. The A and the B_{ba} coefficients are related in the same way as when there is no degeneracy.

(6) The main features of the blackbody spectrum are isotropy (the same in all directions) and homogeneity (the same at all points in the cavity), and the energy density is the same function of frequency independently of the material in the walls of the cavity.

The probability of having n photons is given by

$$p_n = \frac{e^{-n\hbar\omega/kT}}{Z} \tag{13.38}$$

where $Z = \sum_m e^{-m\hbar\omega/kT}$. The energy is given by

$$U = \sum_n p_n n\hbar\omega \tag{13.39}$$

$$= \frac{\hbar\omega}{Z} \sum_n n e^{-n\hbar\omega/kT} \tag{13.40}$$

$$= \hbar\omega \frac{1}{e^{\hbar\omega/kT} - 1} \tag{13.41}$$

In order to obtain Planck's formula we need to multiply this expression by the density of states and this gives the correct result. The terms are: ω (the frequency), c (the speed of light), k (the Boltzmann constant) and T (the temperature).

(i) The average number of photons is

$$\sum_n p_n n = 1/(e^\beta - 1) \tag{13.42}$$

and in order to calculate the deviation we need to calculate $\sum_n p_n n^2$, where $\beta = \hbar\omega/kT$. To do so we can differentiate this expression with respect to β to obtain

$$-\frac{e^\beta}{(e^\beta - 1)^2} = -\langle n^2 \rangle + \langle n \rangle^2 \tag{13.43}$$

which can be rearranged to get

$$\langle n^2 \rangle = \langle n \rangle^2 + \langle n \rangle \frac{e^\beta}{e^\beta - 1} \qquad (13.44)$$

$$= \langle n \rangle^2 + \langle n \rangle (1 + \langle n \rangle) \qquad (13.45)$$

which leads to

$$\langle n^2 \rangle = \langle n \rangle + 2 \langle n \rangle^2 \qquad (13.46)$$

Therefore

$$\sigma^2 = \langle n^2 \rangle - \langle n \rangle^2 = \langle n \rangle (1 + \langle n \rangle) \qquad (13.47)$$

(ii) The part proportional to $\langle n \rangle$ is due to particles and the part proportional to $\langle n^2 \rangle$ reflects the wave behavior. Reason: the standard deviation in the particle number is usually associated with the square root of that number because particles follow some kind of a binomial population distribution (whose standard deviation is $\sqrt{N}$).

13.2   Problem and Solutions 2

13.2.1   *Problem set 2*

(1) An optical wave with angular frequency ω and wave vector k, whose field is

$$E = E_\omega \cos(\omega t - kz) \qquad (13.48)$$

is incident on a second harmonic material with a linear and a second harmonic susceptibility. (i) Write down an expression for the induced polarizability of the medium indicating which components oscillate at which frequency. (ii) The equation characterizing the growth of the second harmonic signal $E_{2\omega}$ is of the form

$$\frac{d}{dz} E_{2\omega} = c E_\omega^2 e^{i\Delta k z} \qquad (13.49)$$

where $\Delta k = k_{2\omega} - 2k_\omega$ is the wave vector mismatch and c is a coefficient which depends on the second order susceptibility of the medium. (a) Find a solution for the growth of the second harmonic field in a crystal of length L; (b) Determine the wave vector mismatch when the refractive indices at the fundamental, chosen to be $500 nm$ and the second harmonic are 1.52 and

1.5201 respectively. (c) For the case in (b), what is the optimum length of the crystal for the maximum second harmonic generation?

(2) The organic dye Rhodamine 6G can lase at wavelengths from 550 to 630nm Estimate the shortest possible pulse duration that can be achieved by mode-locking this material in a laser. How many optical cycles does this represent? If the cavity of the laser has a length $d = 1.5m$, and the average power output is 100mW, what is the peak power of a mode-locked pulse having the shortest duration allowed by the gain medium?

(3) A helium-neon laser consists of mirrors of reflectivity 100 percent at one end, and 99 percent at the other. It has a cavity of gain length 10cm. (i) In terms of the reflectivities, gain etc., what is the minimum gain coefficient α (the gain) for oscillations to be maintained? Evaluate this for the numerical values given. (ii) What is the longitudinal mode spacing for the helium-neon laser? How many of these modes would you expect to oscillate under normal conditions? Take the gain bandwidth to be 1.5GHz.

(4) Explain briefly the classical model of atom light interaction and name its major successes and failures.

   (i) Write down Newton's second law for a driven harmonic oscillator whose natural frequency is ω_0. Assume that the frequency of the driving field is ω. Solve this equation to obtain the amplitude of the oscillator as a function of time. When is the amplitude largest?

   (ii) Write down the modified equation for the oscillator evolution in the presence of damping and explain briefly the origin of radiation damping. Show that the solution of this equation is given by

$$x(t) = e^{-\gamma t/2}(Ae^{i\omega_0 t} + Be^{-i\omega_0 t}) \qquad (13.50)$$

   What is the phenomenon of line broadening? Assume that $A = 0$ and $B = 1$ in the above expression. Using Fourier transforms or otherwise, express the line width (defined at the full width half maximum) as a function of the damping coefficient γ.

   (iii) Estimate (to within factors of π and 2 and so on) the line width due to the Doppler broadening of a hydrogen atomic level with the transition frequency $\omega = 10^{15}Hz$ and at temperature $T = 300K$. State clearly all the assumptions you make in this calculation.

You may use the fact that a Fourier transform of the exponential function $e^{-x/a}$ is given by

$$\Phi(p) = \int_0^\infty e^{-ax} e^{2\pi i p x} dx = \frac{-1}{2\pi i p - a} \tag{13.51}$$

and that the Doppler shift formula reads

$$\omega' = \omega \sqrt{\frac{1 - v/c}{1 + v/c}} \tag{13.52}$$

where c is the speed of light.

(5) Describe briefly the main features of laser light.

  (i) Imagine a one-dimensional vacuum cavity with perfectly reflecting mirrors at $x = 0$ and $x = L$. Write down the expression for the wavelengths and frequencies of the allowed modes of the cavity as a function of L?

  (ii) Suppose that each mode i can be represented as a wave of the form

$$E_i = E_0 e^{i\omega_i t} \tag{13.53}$$

  where the amplitude E_0 is the same for all modes, and ω_i is the frequency of the ith mode. Write down the expression for the total intensity of light inside the cavity taking into account N modes. At what times do the intensity peaks occur and what is the width of each pulse?

  (iii) A more realistic form of the ith mode amplitude is

$$E_i = E_0 e^{i(\omega_i t + \delta_i)} \tag{13.54}$$

  where δ_i is the extra phase of the ith mode. Comment on the physical origin of this extra phase. What is the total intensity as the number of modes becomes very large, assuming that the extra phases, δ_i, are completely randomly distributed?

  (iv) How would you make very short laser pulses with a very high intensity?

(6) Suppose that in our second harmonic generation experiment, we could not fully achieve the phase matching of momenta. What would be the optimal length of the crystal to achieve the maximum in the second harmonic intensity? This is called the coherence length. Compute the coherence length if $\lambda_1 = 10^{-6} m$ and $\lambda_2 = 5 \times 10^{-5} m$ and the corresponding refractive indices are n_1 and n_2.

13.2.2 *Solutions 2*

(1) $P = \epsilon_0(\chi^{(1)}E + \chi^{(2)}E^2 + \cdots)$.

$$E^2 = E_\omega^2 \cos^2(\omega t - kz) = \frac{1}{2}E_\omega^2(1 + \cos 2(\omega t - kz)) \quad (13.55)$$

and

$$P_{2\omega} = \frac{1}{2}\epsilon_0 \chi^{(2)} E_\omega^2 \cos 2(\omega t - kz) \quad (13.56)$$

Therefore

$$P = \frac{1}{2}\epsilon_0 \chi^{(2)} E_\omega^2 + \epsilon_0 \chi^{(1)} E_\omega + P_{2\omega} \quad (13.57)$$

Integrate given equation of motion

$$E_{2\omega} = cE_\omega^2 \int_0^L dz e^{i\Delta kz} \quad (13.58)$$

$$= 2cE_\omega^2 e^{i\Delta kL/2}\frac{\sin \Delta kL/2}{\Delta k} \quad (13.59)$$

Also,

$$\Delta k = k_{2\omega} - k_\omega = \frac{2\omega}{c}(n_{2\omega} - n_\omega) = \frac{4\pi}{\lambda}(n_{2\omega} - n_\omega) \quad (13.60)$$

So, $\Delta k = 2513m^{-1}$. Now the maximum of the second harmonic field is when

$$\sin \Delta k/2 = 1 \quad (13.61)$$

or $L_{\text{opt}} = m\pi/\Delta k$ (m odd).

(2) $\Delta\lambda = 80nm$, $\bar\lambda = 590nm$. Therefore

$$\Delta\nu = \frac{c\Delta\lambda}{\bar\lambda^2} = 6.9 \times 10^{13} Hz \quad (13.62)$$

According to the bandwidth theorem

$$\Delta t \approx \frac{1}{\Delta\nu} = 15fs \quad (13.63)$$

Optical period is

$$T = \frac{\lambda}{c} = 2fs \quad (13.64)$$

Therefore $\Delta t = 7.5$ optical cycles.

If we have mode-locking, than we have a train of pulses separated by the cavity round trip $= 2d/c = 10^{-8}$ seconds. Therefore the pulse repetition rate is 10^8 pulses per second. Hence the power is equal to the pulse energy times 10^8 and is given to be $100mW$. Therefore

$$E_p = 10^{-9}J \qquad (13.65)$$

The peak power is given by

$$\frac{E_P}{\Delta t} = 0.67 \times 10^5 W \qquad (13.66)$$

(3) (i) The threshold for amplification is given by

$$R_1 R_2 e^{\alpha 2l} e^{-\beta 2l} = 1 \qquad (13.67)$$

To get the absolute minimum gain, set all the losses to zero, i.e. $\beta = 0$, so that

$$\alpha_{\min} = \frac{1}{2l} \ln \frac{1}{R_1 R_2} = 0.05 m^{-1} \qquad (13.68)$$

(ii) Mode spacing $= c/2d = 1.9 \times 10^9 Hz$. But gain bandwidth is $1.5 \times 10^9 Hz$, and thus we can fit only one mode inside gain profile. All others see no gain.

(4) In the classical model the electron is treated as an oscillator tied by a spring to a nucleus. Light is now treated as an external force driving the oscillator. This treatment is good at explaining phenomena such as the Rayleigh scattering, but fails when it comes to explaining the blackbody radiation and the Compton scattering.

(i) The second law is

$$\ddot{x} + \omega_0^2 x = \frac{qE}{m} \cos \omega t \qquad (13.69)$$

The general solution is

$$x(t) = A \cos(\omega_0 t - \phi) + B \cos \omega t \qquad (13.70)$$

Substituting this back into the above equation we obtain $B = qE/m(\omega_0^2 - \omega^2)$, so that

$$x = A \cos(\omega_0 t - \phi) + \frac{qE}{m(\omega_0^2 - \omega^2)} \cos \omega t \qquad (13.71)$$

Amplitude is largest when the driving is on resonance.

(ii) The equation of motion including the damping is

$$\frac{d^2 x}{dt^2} + \gamma \frac{dx}{dt} + \omega_0^2 x = 0 \qquad (13.72)$$

We assume that there is no external driving here. The solution is given by

$$x(t) = e^{-\gamma t/2}(Ae^{i\omega_0 t} + Be^{-i\omega_0 t}) \qquad (13.73)$$

This solution exponentially decays to zero.

Let's only look at the negative frequency part of $e^{-\gamma t/2}(Ae^{i\omega_0 t} + Be^{-i\omega_0 t})$. Putting $A = 0$ and $B = 1$ we perform a Fourier transform of it leading to

$$\Phi(\omega) = \frac{1}{i(\omega_0 - \omega) - \gamma/2} \qquad (13.74)$$

The intensity is given by the mod squared

$$I = \frac{1}{(\omega_0 - \omega)^2 + \gamma^2/4} \qquad (13.75)$$

which is a Lorentzian. Now we compute full width half maximum. The maximum of I is when $\omega_0 - \omega$ and is $I_M = 4/\gamma^2$. So we need to ω so that $I = 2/\gamma^2$. This implies

$$(\omega_0 - \omega)^2 = \gamma^2/4 \qquad (13.76)$$

and

$$\omega = \omega_0 \pm \frac{\gamma}{2} \qquad (13.77)$$

Therefore $\Delta\omega = \gamma$.

(iii) The shift in frequency due to the Doppler effect is

$$\omega' = \omega\sqrt{\frac{1 - v/c}{1 + v/c}} \approx \omega\left(1 - \frac{v}{c}\right) \qquad (13.78)$$

We are allowed to approximate since the velocity of atoms at room temperature (about $500m/s$) is much smaller than the velocity of light (about 10^7 times). Therefore

$$\Delta\omega = \omega\frac{v}{c} \qquad (13.79)$$

But according to the equipartition theorem $mv^2/2 = 3kT/2$ so that

$$v = \sqrt{\frac{3kT}{m}} \qquad (13.80)$$

Thus

$$\Delta\omega = \omega\sqrt{\frac{3kT}{mc^2}} \qquad (13.81)$$

Substituting the given values we obtain $\Delta\omega = 9.1 \times 10^9 Hz$.

(5) The name stands for light amplification by stimulated emission. We can think of it as the Bose–Einstein condensate for light: one photon that bounces back and forth in a cavity with two highly reflecting mirrors stimulates two photons into the same state as the original photon. So we get a huge amount of coherent radiation very quickly in this way. If one of the mirrors can also transmit then this output gives us the laser light. Its properties are high intensity, coherence and directionality.

(i) The modes allowed are those whose wavelength satisfies: $n\lambda = 2L$. The corresponding frequency is $\omega_n = 2\pi c/\lambda = \pi c n/L$.

(ii) So we need to calculate

$$E = \sum_{i=1}^{N} E_i = E_0 \sum_i e^{i\omega_i t} \tag{13.82}$$

Suppose that we have N different waves in a cavity. We then have

$$\sum_{n=1}^{N} e^{i\pi nct/L} = \frac{1 - e^{i\pi Nct/L}}{1 - e^{i\pi ct/L}} \tag{13.83}$$

The intensity is

$$I \propto \frac{\sin^2(N\pi ct/2L)}{\sin^2(\pi ct/2L)} \tag{13.84}$$

The peaks occur when $t = 2Ln/c$. The width of each pulse is given by $2L/cN$ (distance from the central maximum at $t = 0$ to the first minimum).

(iii) The extra phases come from the fact that the atoms in the walls of the cavity emit photons at random times. The intensity is now

$$I = E^* E = \sum_{jk} |E_0|^2 e^{-i\phi_j} e^{i\phi_k} \tag{13.85}$$

$$= |E_0|^2 (N + \sum_{j\neq k} e^{i(\phi_k - \phi_j)}) \tag{13.86}$$

$$= |E_0|^2 (N + 2\sum_{j>k} \cos((\phi_k - \phi_j))) \tag{13.87}$$

The average of cos is zero since phases are random so that

$$\langle I \rangle = |E_0|^2 N \qquad (13.88)$$

(iv) The pulses can be made short by making sure that the extra phase of as many modes as possible are the same (known as mode-locking) and this would also lead to an increase in intensity which would then be proportional to N^2.

(6) We have that $\Delta k L = \pi/2$. Now, $\delta k = 2n_1 2\pi/10^{-6} - 2\pi n_2/5 \times 10^{-5}$. Therefore, since $n_1 = 1.509$ and $n_2 = 1.530$, we have that $L = 1.210^{-5}$.

13.3 Problems and Solutions 3

13.3.1 *Problem set 3*

(1) Suppose that a two level system is in the state $a|0\rangle + b|1\rangle$. What are the probabilities of observing 0 and 1? What are the probabilities of observing $|\pm\rangle = (|0\rangle \pm |1\rangle)/\sqrt{2}$?

(2) A two state system with energy separation between the two states of $\hbar\omega_0$ is prepared in its ground state at $t = 0$. At $t = 0$ a sinusoidal perturbation

$$V(t) = \hbar V_0 \cos \omega t \qquad (13.89)$$

is applied to the system (this assumes that there are only the off-diagonal elements, so that $V_{aa} = V_{bb} = 0$). Solving the time dependent Schrödinger equation in the two state basis and neglecting rapidly varying terms, show that the maximum probability of being in the excited state is given by

$$P_{\text{max}} = \frac{|V_0|^2}{(\omega - \omega_0)^2 + |V_0|^2} \qquad (13.90)$$

where $|V_0|$ is the matrix element of V_0 between the two states of the system. Show that a pulsed perturbation of duration $t = \pi/|V_0|$ with frequency $\omega = \omega_0$ inverts the system so that $p = 1$.

(3) A three level system interacting with an electromagnetic field can be described by the following Hamiltonian:

$$H_0 = \hbar\omega_a |a\rangle\langle a| + \hbar\omega_b |b\rangle\langle b| + \hbar\omega_c |c\rangle\langle c| \qquad (13.91)$$

$$V(t) = -\frac{\hbar}{2}(\Omega_1 e^{-i(\phi_1 + \nu_1)t}|a\rangle\langle b| + \Omega_2 e^{-i(\phi_2 + \nu_2)t}|a\rangle\langle c|) + \text{h.c.}$$

Write down the total Hamiltonian in the matrix form. Suppose first that there is no interaction. What is the free evolution of the system? Suppose then that the interaction is turned on and that $\phi_1 = -\nu_1$ and $\phi_2 = -\nu_2$. Find the eigenvectors of the resulting interaction Hamiltonian.

(4) Explain in detail the semi-classical model of light matter interactions. Name one deficiency of this model and explain your choice.

A non-interacting two level atom is described by the following Hamiltonian:

$$H_0 = E_1|1\rangle\langle1| + E_2|2\rangle\langle2| \tag{13.92}$$

Explain the physical meaning of each term in this expression. This atom then interacts with the electromagnetic field such that the interaction Hamiltonian is given by

$$V(t) = \gamma e^{i\omega t}|1\rangle\langle2| + \gamma e^{-i\omega t}|1\rangle\langle2| \tag{13.93}$$

Explain the physical meaning of every term in the interaction Hamiltonian.

Assume that the atom is initially in the state 1. By knowing that the state of the atom is in general given by

$$c_1(t)e^{-iE_1t}|1\rangle + c_2(t)e^{-iE_1t}|1\rangle \tag{13.94}$$

solve the Schrödinger equation to show that the probability of occupying the exited state oscillates at the frequency

$$\Omega = \sqrt{\gamma^2/\hbar^2 + (\omega - \omega_{12})^2/4)} \tag{13.95}$$

i.e.

$$|c_2(t)|^2 \propto \sin^2 \Omega t \tag{13.96}$$

where

$$\omega_{12} = \frac{E_2 - E_1}{\hbar} \tag{13.97}$$

(Do not derive the constant of proportionality.)

Assume that the interaction is "on-resonant". Derive and plot the exact evolution of the probability to occupy the excited state 2 as a function of time. Given that $\gamma = 10^{-24}J$, how long does it take to excite the atom to level 2 if it is initially in level 1?

(5) Explain briefly the semi-classical approximation in the treatment of light matter interactions.

A nuclear spin has two possible states in a external magnetic field, up $|\uparrow\rangle$ (i.e. aligned with the field) and down $|\downarrow\rangle$ (i.e. anti-aligned with the field). Suppose that the nucleus is in an external (static) magnetic field of strength B, which points in the z direction.

   (i) Write down the Hamiltonian for the nucleus using the Pauli matrix notation and identify its eigenvalues.
   (ii) Suppose that the initial state of the system is aligned with the field in the z direction, $|\uparrow\rangle$. Suppose then that the field is instantaneously switched to the x direction. Solve the Schrödinger equation to obtain the exact evolution of the nuclear spin in terms of the eigenstate of the Pauli spin matrix σ_x. What is the phase difference between the two orthogonal spin eigenstates of σ_x as a function of time?
   (iii) After what time will the spin switch to *its* orthogonal state $|\downarrow\rangle$?

What is the relationship between the energy associated with the spin and the time it takes to evolve between orthogonal states? Comment on the validity of the time energy uncertainty relation to estimate the time of this transition.

The Pauli matrices are given in the $|\uparrow\rangle$, $|\downarrow\rangle$ basis by

$$\sigma_x = \begin{pmatrix} 0 & 1 \\ 1 & 0 \end{pmatrix}, \sigma_y = \begin{pmatrix} 0 & -i \\ i & 0 \end{pmatrix}, \sigma_z = \begin{pmatrix} 1 & 0 \\ 0 & -1 \end{pmatrix} \tag{13.98}$$

13.3.2 *Solutions 3*

(1) The probabilities in the $0, 1$ basis are $|a|^2$ and $|b|^2$, while in the $\pm$ basis they are $|a + b|^2$ and $|a - b|^2$.

(2) Let's start by writing

$$\Psi = c_1(t)e^{-iE_1 t/\hbar}|1\rangle + c_2(t)e^{-iE_2 t/\hbar}|2\rangle \tag{13.99}$$

Substituting this into the Schrödinger equation

$$i\hbar\frac{\partial\Psi}{\partial t} = (H_0 + \hbar V(t))\Psi \tag{13.100}$$

we obtain

$$\dot{c}_1 = -i\bar{V}_0 \cos\omega t e^{-i\omega_0 t} c_2 \tag{13.101}$$
$$\dot{c}_2 = -i\bar{V}_0^* \cos\omega t e^{i\omega_0 t} c_1 \tag{13.102}$$

After applying the rotating wave approximation we obtain

$$\dot{c}_1 = -i\bar{V}_0 e^{i(\omega - \omega_0)t} c_2 \qquad (13.103)$$

$$\dot{c}_2 = -i\bar{V}_0^* e^{-i(\omega - \omega_0)t} c_1 \qquad (13.104)$$

The initial conditions are $c_1(0) = 1, c_2(0) = 0$ and $\dot{c}_2(0) = -i/2\bar{V}_0^*, \dot{c}_1(0) = 0$. Eliminating c_1 by double differentiation we arrive at

$$\ddot{c}_2 + i(\omega - \omega_0)\dot{c}_2 + \frac{1}{4}|\bar{V}_0|^2 c_2 = 0 \qquad (13.105)$$

A trial solution is $e^{i\mu t}$ and the resulting equation is

$$-\mu^2 - (\omega - \omega_0)\mu + \frac{1}{4}|\bar{V}_0|^2 = 0 \qquad (13.106)$$

The roots are

$$\mu_\pm = \frac{1}{2}(-(\omega - \omega_0) \pm [(\omega - \omega_0)^2 + |\bar{V}_0|^2]^{1/2}) \qquad (13.107)$$

and the full solution to the Schrödinger equation is thus

$$c_2(t) = A_+ e^{i\mu_+ t} + A_- e^{i\mu_- t} \qquad (13.108)$$

The coefficients $A_\pm$ can be fixed from the initial conditions to yield

$$c_2(t) = -\frac{|\bar{V}_0|^2}{2} \frac{1}{(\omega - \omega_0)^2 + |\bar{V}_0|^2}(e^{i\mu_+ t} - e^{i\mu_- t}) \qquad (13.109)$$

where

$$\Omega = (\omega - \omega_0)^2 + |\bar{V}_0|^2]^{1/2} \qquad (13.110)$$

is the Rabi frequency. Thus the solution for the upper level evolution is

$$c_2(t) = -\frac{i}{\Omega}\bar{V}_0^* e^{-i(\omega - \omega_0)t/2} \sin\frac{1}{2}\Omega t \qquad (13.111)$$

The probability in question now follows from this. The flip is reached when $\sin\frac{1}{2}\Omega t = 1$, which results in the π-pulse of duration π/Ω.

(3) Solve the Schrödinger equation to obtain the free evolution. The solution should be of the form

$$|\Psi(t)\rangle = c_a e^{-i\omega_a t}|a\rangle + c_b e^{-i\omega_b t}|b\rangle + c_c e^{-i\omega_c t}|c\rangle \qquad (13.112)$$

where c_a, c_b, c_c are initial amplitudes for states a, b, c respectively. When you diagonalize the interaction Hamiltonian you should obtain the following eigenstates:

$$|\psi_\pm\rangle = \frac{1}{\sqrt{2}}\left(|a\rangle \pm \frac{\Omega_1}{\Omega}|b\rangle \pm \frac{\Omega_2}{\Omega}|c\rangle\right) \qquad (13.113)$$

$$|\psi_0\rangle = \left(\frac{\Omega_2}{\Omega}|b\rangle - \frac{\Omega_1}{\Omega}|c\rangle\right) \qquad (13.114)$$

where $\Omega = \sqrt{\Omega_1^2 + \Omega_2^2}$. (What are the corresponding eigenvalues?).

(4) In the semi-classical model atoms are quantized, but light is not. The atom is described through a Hamiltonian, and the effect of light is taken as an additional part of the Hamiltonian. The evolution of the system is obtained by solving the Schrödinger equation with the total Hamiltonian. This is frequently impossible to solve analytically and we have to resort to approximations or numerics. The effect of oscillating fields is usually taken as a perturbation of the basic non-interacting atomic Hamiltonian. This leads to the time dependent perturbation theory where the most useful result is Fermi's golden rule. This tells us the probability of obtaining a transition from one level to another under a time dependent perturbation.

$|1\rangle$ and $|2\rangle$ represent the two atomic levels. The E_1 and E_2 are the corresponding energies of the two states. They are the eigenvalues of the atomic Hamiltonian with $|1\rangle$ and $|2\rangle$ being the eigenvectors. Now, when this atom interacts with a field the Hamiltonian contains the transition elements for jumping from 1 to 2 and vice versa. We have the creation and annihilation operators $|2\rangle\langle 1|$ and $|1\rangle\langle 2|$.

The Schrödinger equation is

$$(E_1|1\rangle\langle 1| + E_2|2\rangle\langle 2| + \gamma e^{i\omega t}|1\rangle\langle 2| + \gamma e^{-i\omega t}|2\rangle\langle 1|)|\Psi(t)\rangle$$
$$= i\hbar\frac{\partial|\Psi(t)\rangle}{\partial t} \qquad (13.115)$$

By substituting in the wave function

$$|\Psi(t)\rangle = c_1(t)e^{-iE_1 t}|1\rangle + c_2(t)e^{-iE_2 t}|2\rangle \qquad (13.116)$$

we obtain two equations:

$$\frac{\gamma}{\hbar}c_2 = i\dot{c}_1 e^{i(\omega_{12}-\omega)t} \qquad (13.117)$$

$$\frac{\gamma}{\hbar}c_1 = i\dot{c}_2 e^{-i(\omega_{12}-\omega)t} \qquad (13.118)$$

This is a system of coupled equations which we solve for c_2 by differentiating the second equation and substituting the first equation into it. We obtain

$$\ddot{c}_2 - i(\omega_{12} - \omega)\dot{c}_2 + \frac{\gamma^2}{\hbar^2}c_2 = 0 \qquad (13.119)$$

The trial solution $c_2 = e^{i\mu t}$ leads to

$$\mu^2 - (\omega_{12} - \omega)\mu - \frac{\gamma^2}{\hbar^2} = 0 \qquad (13.120)$$

which has two roots:

$$\mu_{1,2} = \frac{\omega_{12} - \omega}{2} \pm \left(\sqrt{\gamma^2/\hbar^2 + (\omega - \omega_{12})^2/4}\right) \qquad (13.121)$$

Therefore

$$c_2(t) = Ae^{i\mu_1 t} + Be^{i\mu_2 t} \qquad (13.122)$$

But $c_2(0) = 0$, thus $A = -B$. The solution is therefore

$$|c_2|^2 = 4A^2 \sin^2(\Omega t) \qquad (13.123)$$

as required.

If on resonance, we have

$$|c_2|^2 = \sin^2(\gamma t/\hbar) \qquad (13.124)$$

For $|c_2|^2 = 1$ we require

$$\frac{\gamma t}{\hbar} = \frac{\pi}{2} \qquad (13.125)$$

and so we obtain the time for a flop to be $t = \pi\hbar/2\gamma$. Given that $\gamma = 10^{-24}$J, we find $t = 1.65 \times 10^{-10}$ seconds.

(5) The first part is the same as the previous question.

  (i) The initial Hamiltonian is $H = \frac{\mu B}{2}\sigma_z$. This has eigenvalues $-\mu B/2$ and $+\mu B/2$ as can be seen from the matrix form:

$$H = \begin{pmatrix} \frac{\mu B}{2} & 0 \\ 0 & -\frac{\mu B}{2} \end{pmatrix}$$

  (ii) Now the state is $|\uparrow\rangle$, but the Hamiltonian is $H = \frac{\mu B}{2}\sigma_x$. The eigenvectors of this Hamiltonian are $|\rightarrow\rangle$ and $|\leftarrow\rangle$, which evolve with phases $e^{-i\mu Bt/2\hbar}$ and $e^{i\mu Bt/2\hbar}$ respectively. But, the initial state can be written as an equal superposition of the eigenstates of σ_x:

$$|\uparrow\rangle = |\leftarrow\rangle + |\rightarrow\rangle \qquad (13.126)$$

Therefore the state after some time T is given by

$$|\psi(T)\rangle = e^{i\mu BT/2\hbar}|\leftarrow\rangle + e^{-i\mu BT/2\hbar}|\rightarrow\rangle \qquad (13.127)$$

The phase difference between the two states is

$$\Delta\phi = \frac{\mu BT}{\hbar} \qquad (13.128)$$

(3) The orthogonal state to the initial state is

$$|\downarrow\rangle = |\leftarrow\rangle - |\rightarrow\rangle \qquad (13.129)$$

So, the phase difference between the states has to be $e^{i\mu BT/\hbar} = -1$ and so

$$T = \frac{\pi\hbar}{\mu B} \qquad (13.130)$$

The energy difference is $E = \mu B$ and so from the last question we have that

$$TE = \pi\hbar \qquad (13.131)$$

This is very similar to the uncertainty relation between energy and time $\Delta E \Delta t \geq \hbar$.

Thus the time energy uncertainty is a pretty good estimate of the time it takes for the transition, given that it is only out by π.

13.4   Problems and Solutions 4

13.4.1   *Problem set 4*

(1) In the Schrödinger picture, operators are time independent and states evolve according to the Schrödinger equation

$$i\hbar\frac{\partial}{\partial t}|\Psi\rangle = \hat{H}|\Psi\rangle \qquad (13.132)$$

where $\hat{H}$ is the (time independent) Hamiltonian. Find a form for the time evolution operator

$$\hat{U}(t,t_0)|\Psi(t_0)\rangle = |\Psi(t)\rangle \qquad (13.133)$$

in terms of the Hamiltonian $\hat{H}$. Use this to show that in the Heisenberg picture, where states do not evolve in time but operators representing observables do, that

$$i\hbar\frac{d}{dt}\hat{O}(t) = [\hat{O},\hat{H}] \qquad (13.134)$$

(2) Consider a two level atom with the ground state $|g\rangle$ and excited state $|e\rangle$ interacting with a single quantized radiation field mode of frequency ω which is close (but not equal) to the atomic transition frequency ω_0. The interaction energy between the atomic dipole $\hat{d}$ and the electric field $\hat{E}$ is $\hat{V} = -\hat{d}\hat{E}$. We write the dipole operator as $\hat{d} = d_{eg}(\sigma_+ + \sigma_-)$ where $\sigma_\pm$ are the Pauli raising and lowering operators

$$\hat{\sigma}_+|g\rangle = |e\rangle \qquad (13.135)$$

$$\hat{\sigma}_-|e\rangle = |g\rangle \qquad (13.136)$$

so $\hat{\sigma}_+ = |e\rangle\langle g|$. The field operator $\hat{E} = E_0(\hat{a} + \hat{a}^\dagger)\sin kz$, where $\hat{a}$ and $\hat{a}^\dagger$ are the creation and annihilation operators. Use the result of question 1 to show that the rotating wave approximation of neglecting fast varying non-resonant coupling is equivalent to dropping terms $\hat{a}^\dagger\hat{\sigma}_- + \hat{a}\hat{\sigma}_+$ in V.

(3) A single mode field is prepared in a number state $|n\rangle$ with precisely n photons. Calculate the uncertainty in the field operator

$$\hat{E} = E_0(\hat{a} + \hat{a}^\dagger)\sin kz \qquad (13.137)$$

and interpret the result.

(4) Explain the process of quantizing the electromagnetic field. Why is a single mode of the field equivalent to a unit mass harmonic oscillator?

In terms of the creation and annihilation operators, the Hamiltonian for a single mode field of frequency ω is

$$H = \hbar\omega\left(\hat{a}^\dagger\hat{a} + \frac{1}{2}\right) \qquad (13.138)$$

A coherent state of this field with the amplitude α is given by

$$|\alpha\rangle = e^{-|\alpha|^2/2}\sum_n \frac{\alpha^n}{\sqrt{n!}}|n\rangle \qquad (13.139)$$

What is the probability of obtaining n photons in the field? What is the average energy of the field in this state? Hence give the physical meaning of the amplitude α.

Solve the Schrödinger equation to obtain the free evolution of this state. What happens to the amplitude α during the evolution?

The action of a beam splitter with the transmission and reflection amplitudes T and R respectively is given by

$$|n\rangle \otimes |0\rangle \rightarrow \sum_p T^{n-p} R^p \sqrt{\binom{n}{p}} |n-p\rangle \otimes |p\rangle \qquad (13.140)$$

Suppose that the input is a coherent state of amplitude α in one port and vacuum in the other port, $|\alpha\rangle \otimes |0\rangle$. Show that the output is a product of coherent states

$$|T\alpha\rangle \otimes |R\alpha\rangle \qquad (13.141)$$

Hence explain why the coherent state is considered the best quantum description of classical light.

(5) A quantum particle moving non-relativistically in one dimension has mass m and potential energy $\frac{1}{2}m\omega^2 x^2$. Write down its Hamiltonian H. Express H in terms of the operators

$$a = \frac{\beta}{\sqrt{2}}\left(x + i\frac{p}{m\omega}\right) \qquad (13.142)$$

$$a^\dagger = \frac{\beta}{\sqrt{2}}\left(x - i\frac{p}{m\omega}\right) \qquad (13.143)$$

where $\beta^2 = m\omega/\hbar$. You may assume that $[x,p] = i\hbar$.

(i) Evaluate the commutators $[a, a^\dagger]$, $[H, a^\dagger]$ and $[H, a]$.
(ii) Hence determine the allowed energy levels of the particle, explaining carefully the logic that you use. What do these levels represent when we apply them to a single mode of the quantized electromagnetic field?
(iii) Let $|0\rangle$ denote the ground state. Show that

$$\langle 0|(a + a^\dagger)|0\rangle = 0 \qquad (13.144)$$
$$\langle 0|(a + a^\dagger)^2|0\rangle = 1 \qquad (13.145)$$

What do these relationships signify in relation to the quantized electromagnetic field?

(6) Suppose that we have a state of two light modes of the form $1/2|2,0\rangle + 1/\sqrt{2}|1,1\rangle + 1/2|2,0\rangle$. Prove that this state is entangled, i.e. prove that the state cannot be written as a state of light $(a|0\rangle + b|1\rangle + c|2\rangle)(e|0\rangle + f|1\rangle + g|2\rangle)$.

13.4.2 *Solutions 4*

(1) From the Schrödinger equation it follows that

$$|\Psi(t)\rangle = e^{-i\hat{H}t/\hbar}|\Psi(0)\rangle \qquad (13.146)$$

(Note that this is only true if the Hamiltonian is time independent.) Therefore

$$|\Psi(t)\rangle = e^{-i\hat{H}(t-t_0)/\hbar}|\Psi(t_0)\rangle \qquad (13.147)$$

and so $\hat{U}(t,t_0) = e^{-i\hat{H}(t-t_0)/\hbar}$ is the time evolution operator. The usual expectation value of an observable is given by

$$\langle \hat{O}(t)\rangle = \langle \Psi(t)|\hat{O}|\Psi(t)\rangle \qquad (13.148)$$

But this can be rewritten using our time evolution operator as

$$\langle \hat{O}(t)\rangle = \langle \Psi(0)|e^{i\hat{H}(t-t_0)/\hbar}\hat{O}e^{-i\hat{H}(t-t_0)/\hbar}|\Psi(0)\rangle \qquad (13.149)$$

and so

$$\hat{O}(t,t_0) = e^{i\hat{H}(t-t_0)/\hbar}\hat{O}e^{-i\hat{H}(t-t_0)/\hbar} \qquad (13.150)$$

is the time dependent operator in the Heisenberg picture. Let's now derive the evolution equation for the above operator. Differentiate the equation obtaining

$$i\hbar\frac{d}{dt}\hat{O}(t,t_0) = -\hat{H}\hat{O}(t) + \hat{O}(t)\hat{H} = [\hat{O}(t),\hat{H}] \qquad (13.151)$$

(2) Using the previous question we have

$$i\hbar\frac{d}{dt}\sigma_+ = [\hat{\sigma}_+,\hat{H}] \qquad (13.152)$$

where $\hat{H} = \hat{H}_{\text{atom}} + \hat{H}_{\text{field}}$ and $[\hat{\sigma},\hat{H}_F] = 0$. The trick here is to identify the Hamiltonian for the atom. It is most conveniently given by

$$\hat{H}_{\text{atom}} = E_e|e\rangle\langle e| + E_g|g\rangle\langle g| \qquad (13.153)$$

Let's assume for simplicity that $E_g = 0$. Then $E_e = \hbar\omega_0$ which is the transition energy. So

$$\frac{d}{dt}\sigma_+ = -\frac{i}{\hbar}[\sigma_+, \hbar\omega_0|e\rangle\langle e|] \qquad (13.154)$$

$$= -\frac{i}{\hbar}(\hbar\omega_0[|e\rangle\langle g|e\rangle\langle e| - |e\rangle\langle e|e\rangle\langle g|]) \qquad (13.155)$$

so that

$$\frac{d}{dt}\sigma_+ = \frac{i}{\hbar}(\hbar\omega_0)\sigma_+ \qquad (13.156)$$

$$\sigma_+(t) = e^{i\omega_0 t}\sigma_+(0) \qquad (13.157)$$

Similarly,

$$\sigma_-(t) = e^{-i\omega_0 t}\sigma_-(0) \qquad (13.158)$$

Now we do the same for the field equations. We use $\hat{H}_F = \hbar\omega(\hat{a}^\dagger\hat{a} + 1/2)$ and $[\hat{a}, \hat{a}^\dagger] = 1$. Then

$$i\hbar\frac{d}{dt}\hat{a}(t) = [\hat{a}(t), \hat{H}] = [\hat{a}(t), \hat{H}_F] \qquad (13.159)$$

is the evolution equation which yields (prove it!)

$$\hat{a}(t) = \hat{a}(0)e^{-i\omega t} \qquad (13.160)$$
$$\hat{a}^\dagger(t) = \hat{a}^\dagger(0)e^{i\omega t} \qquad (13.161)$$

So we can write the interaction as

$$\hat{V}(t) = -\hat{d}\hat{E}(t) = -d_{eg}(\hat{\sigma}_+ + \hat{\sigma}_-)E_0\sin kz(\hat{a}^\dagger(t) + \hat{a}(t)) \qquad (13.162)$$

Terms $\sigma_+a \sim e^{i(\omega-\omega_0)t}$ and $\sigma_-a^\dagger \sim e^{-i(\omega-\omega_0)t}$ are slowly varying near resonance, but $\sigma_+a^\dagger \sim e^{i(\omega+\omega_0)t}$ and $\sigma_-a \sim e^{-i(\omega+\omega_0)t}$ very at $\sim 2\omega_0$ and average to zero, leading to the rotating wave approximation.

(3) The mean field $\langle E \rangle$ is zero in a number state, as

$$\langle n|a + a^\dagger|n \rangle = 0 \qquad (13.163)$$

On the other hand, the mean square field is

$$\langle n|E^2|n \rangle = E_0^2\sin^2 kz\langle n|a^2 + a^\dagger a + aa^\dagger + (a^\dagger)^2|n \rangle$$
$$= E_0^2\sin^2 kz(2n + 1) \qquad (13.164)$$

The field with fixed n has definite amplitude, but random phase between 0 and 2π.

(4) The electromagnetic field is classically a wave described by six numbers at every point in space and time. These numbers can be specified simultaneously and the values of both the electric and magnetic field can in principle be determined exactly and simultaneously. When the field is quantized this is no longer possible. In fact, the electric and magnetic fields become operators which are no longer commuting, i.e. they are no longer simultaneously measurable. Writing the total classical energy in the field we get

$$W = \frac{1}{2}\int(\epsilon E^2 + \mu H^2)dV \qquad (13.165)$$

But $E \propto x\cos\omega t/V$ and $B \propto p\omega\sin\omega t/V$, so that $W = (p^2 + \omega^2 x^2)/2$, which is a simple harmonic oscillator.

In terms of the creation and annihilation operators, the Hamiltonian for a single mode field of frequency ω is

$$H = \hbar\omega \left(\hat{a}^\dagger \hat{a} + \frac{1}{2} \right) \qquad (13.166)$$

A coherent state of this field with the amplitude α is given by

$$|\alpha\rangle = e^{-|\alpha|^2/2} \sum_n \frac{\alpha^n}{\sqrt{n!}} |n\rangle \qquad (13.167)$$

What is the probability of obtaining n photons in the field? The probability of obtaining n photons is $|\langle n|\alpha\rangle|^2$ and is

$$p_n = e^{-|\alpha|^2} \frac{|\alpha|^{2n}}{n!} \qquad (13.168)$$

The average energy is

$$\langle E \rangle = \hbar\omega \langle \alpha | \left(\hat{a}^\dagger \hat{a} + \frac{1}{2} \right) |\alpha\rangle \qquad (13.169)$$

$$= \hbar\omega \left(e^{-|\alpha|^2} \sum_n \frac{|\alpha|^{2n}}{n!} n + \frac{1}{2} \right) \qquad (13.170)$$

$$= \hbar\omega \left(e^{-|\alpha|^2} |\alpha|^2 \sum_n \frac{|\alpha|^{2(n-1)}}{(n-1)!} + \frac{1}{2} \right) \qquad (13.171)$$

$$= \hbar\omega \left(|\alpha|^2 + \frac{1}{2} \right) \qquad (13.172)$$

The physical meaning of $|\alpha|^2$ is that it is the average number of photons.

We now solve the Schrödinger equation to obtain the free evolution of this state,

$$i\hbar \frac{\partial |n\rangle}{\partial t} = \hbar\omega \left(\hat{a}^\dagger \hat{a} + \frac{1}{2} \right) |n\rangle \qquad (13.173)$$

The solution is

$$|n(t)\rangle = e^{-i\omega(n+1/2)t} |n\rangle \qquad (13.174)$$

Therefore

$$|\alpha(t)\rangle = e^{-|\alpha|^2/2} \sum_n \frac{\alpha^n}{\sqrt{n!}} e^{-i\omega(n+1/2)t} |n\rangle \qquad (13.175)$$

$$= e^{-i\omega t/2} e^{-|\alpha|^2/2} \sum_n \frac{(\alpha e^{-i\omega t})^n}{\sqrt{n!}} |n\rangle \qquad (13.176)$$

$$= e^{-i\omega t/2} |\alpha e^{-i\omega t}\rangle \qquad (13.177)$$

Therefore the amplitude oscillates at frequency ω.

When the coherent state is the input we have

$$|\alpha\rangle \otimes |0\rangle \rightarrow e^{-|\alpha|^2/2} \sum_n \sum_p T^{n-p} R^p \sqrt{\binom{n}{p}} \frac{\alpha^n}{\sqrt{n!}} |n-p\rangle \otimes |p\rangle$$

$$= e^{-|T\alpha|^2/2} e^{-|R\alpha|^2/2} \sum_n \sum_p \left\{ \frac{T\alpha^{n-p}}{\sqrt{(n-p)!}} \right\} |n-p\rangle$$

$$\otimes \left\{ \frac{R\alpha^p}{\sqrt{p!}} \right\} |p\rangle$$

$$= |T\alpha\rangle \otimes |R\alpha\rangle \tag{13.178}$$

QED.

(5) The Hamiltonian is given by

$$H = \left(\frac{p^2}{2m} + \frac{m\omega^2 x^2}{2} \right) \tag{13.179}$$

We can first compute

$$a^\dagger a = \frac{\beta^2}{2} \left(x^2 + \frac{i}{m\omega}[x,p] + \frac{p^2}{m\omega} \right) \tag{13.180}$$

Using the fact that $[x,p] = i\hbar$ we can rearrange the above to obtain

$$H = \hbar\omega \left(a^\dagger a + \frac{1}{2} \right) \tag{13.181}$$

(i) We can easily prove that $[a, a^\dagger] = 1$, from which it follows that

$$[H, a] = \hbar\omega[a^\dagger, a]a = -\hbar\omega a \tag{13.182}$$

and

$$[H, a^\dagger] = \hbar\omega a^\dagger[a, a^\dagger] = \hbar\omega a^\dagger \tag{13.183}$$

(ii) Suppose that $H|\psi_n\rangle = E_n|\psi_n\rangle$. Then $(Ha - aH)|\psi_n\rangle = (H - E_n)a|\psi_n\rangle = -\hbar\omega a|\psi_n\rangle$. Thus,

$$H(a|\psi_n\rangle) = (E_n - \hbar\omega)(a|\psi_n\rangle) \tag{13.184}$$

Thus, a lowers energy by $\hbar\omega$ and, likewise, it can be shown that $a^\dagger$ raises energy by $\hbar\omega$.

The lowest energy state is the one which cannot be lowered anymore, so

$$a|0\rangle = 0 \tag{13.185}$$

Therefore, $H|0\rangle = \hbar\omega/2|0\rangle$ and so the energy of the nth eigenstate is $E_n = (n + 1/2)\hbar\omega$. When we quantize the electromagnetic field, then each mode (frequency in other words) is a harmonic oscillator. Different levels represent different number of photons that can exist at the given frequency (or with the corresponding energy).

(iii) The first equality is easy as $a|0\rangle = 0$ and likewise $\langle 0|a^\dagger = 0$. The second is done by expansion, since

$$(a + a^\dagger)^2 = a^2 + aa^\dagger + a^\dagger a + (a^\dagger)^2 \qquad (13.186)$$

All terms disappear apart from

$$\langle 0|aa^\dagger|0\rangle = 1 \qquad (13.187)$$

The first relationship says that the average of the electric field in the vacuum is zero. The second says that the average of the energy squared is non-zero. This is a purely quantum signature, and (loosely speaking) says that the vacuum state also contains some energy, since energy of the field is proportional to its square.

(6) It is clear that there are no coefficients a, b, c, d, e, f and g that can reproduce the original state. Therefore the state cannot be written as a product and this means that it is entangled.

13.5 Problems and Solutions 5

13.5.1 *Problem set 5*

(1) The coherent state $|\alpha\rangle$ is generated from the vacuum state of a field mode $|0\rangle$ by a unitary transformation:

$$|\alpha\rangle = \hat{D}(\alpha)|0\rangle \qquad (13.188)$$

where $\hat{D}(\alpha) = e^{+\alpha a^\dagger}e^{-|\alpha|^2/2}$ is the Glauber displacement operator, $\hat{a}^\dagger$ the creation operator. Show that, using this definition, the probability of finding n photons in a coherent state $|\alpha\rangle$, $p(n) = |\langle n|\alpha\rangle|^2$ is given by the Poisson formula

$$p(n) = \frac{\bar{n}^n e^{-\bar{n}}}{n!} \qquad (13.189)$$

and determine the mean number of photons $\bar{n}$ and the standard deviation $\Delta n = [\langle n^2\rangle - \langle n\rangle^2]^{1/2}$.

Calculate the mean field $\langle \hat{E} \rangle$ for an electric field operator $\hat{E} = E_0 \sin kz(\hat{a} + \hat{a}^\dagger)$ for a field prepared in a number state. Then calculate the mean square field $\langle \hat{E}^2 \rangle$ in such a coherent state, and from these results find the field uncertainty $\Delta E = [\langle \hat{E}^2 \rangle - \langle \hat{E} \rangle^2]^{1/2}$. How does this compare with the field uncertainty for the vacuum?

(2) Imagine we can prepare a single field mode in a superposition of zero and ten photons

$$|\Psi\rangle = a|0\rangle + b|10\rangle \qquad (13.190)$$

Take $a = b = 1/\sqrt{2}$ for simplicity. Calculate the mean photon number for this superposition. Suppose having prepared initially such a superposition state, we monitor the field and detect a single photon which has leaked out of the cavity confirming the field mode. What is the field state inferred after such a detection? What is the field mode mean photon number inferred immediately after this leaking photon has been detected? Interpret this surprising result.

(3) Show that the ground state when a cavity field interacting with a two level atom is far detuned from resonance by Δ is Stark shifted in energy by interaction by an amount

$$\delta E = \frac{|\langle 1|V|2\rangle|^2}{\Delta \hbar} \qquad (13.191)$$

(4) Explain why a coherent state is a good mathematical representation of typical laser light. Describe briefly the basic fully quantum mechanical description of light matter interaction.

A two level atom interacts on resonance with a single mode light field. Suppose that the atom is initially excited and the field has n photons. The atom field interaction is described by the Jaynes–Cummings Hamiltonian

$$H = \hbar \lambda (\sigma_- a^\dagger + \sigma_+ a) \qquad (13.192)$$

Explain the physical significance of this Hamiltonian and the meaning of all the symbols.

Prove that the joint state of the atom and field at some time t is given by

$$\Psi(t) = \cos \lambda_n t |e, n\rangle + i \sin \lambda_n t |g, n+1\rangle \qquad (13.193)$$

Give the expression for λ_n.

Calculate the probability for the atom to be in the ground state and plot it as a function of time.

Now suppose that the field is initially in a coherent state of amplitude α instead of a number state. Using linearity of Schrödinger's equation and *without performing any calculation whatsoever* write down the expression for the probability that after time t the atom is in the ground state.

Assume that $\alpha = 0$. After what time is this probability greater than a half?

(5) An atom has two energy levels, i and j, separated in energy by $\hbar\omega_{ij}$. It is subject to a small external time dependent monochromatic electromagnetic perturbation for a time T, oscillating at the frequency ω. You may assume that the perturbation has matrix elements $V_{ji} = V_{ij}^*$ between these states. Show that if the atom is initially in the state i, the probability of a transition to the state j is approximately

$$P_{ij} = 4|V_{ij}|^2 \frac{\sin^2((\omega_{ij} - \omega)T/2)}{(\hbar(\omega_{ij} - \omega))^2} \qquad (13.194)$$

   (i) Argue that the probability of the transition back from j to i is the same as that for the transition from i to j.
   (ii) Show that, within the formalism employed, the transition rate grows linearly with time.
   (iii) Why is the Einstein B coefficient independent of time (only a qualitative explanation required)?

(6) Suppose that we have the following two level atom and single filed mode interaction Hamiltonian:

$$H = (\sigma_-(a^\dagger)^4 + \sigma_+ a^4) \qquad (13.195)$$

Compute the dynamics of the initial state $|e, 0\rangle$.

13.5.2 *Solutions 5*

(1) Here we have to compute $\langle n|D(\alpha)|0\rangle$. Therefore we need to be able to evaluate

$$\langle n|e^{\alpha a^\dagger}|0\rangle \qquad (13.196)$$

To do so, you can use the fact that $e^x = \sum_n x^n/n!$, and then apply the creation operator algebra. The rest of the question can be answered straightforwardly from the relevant chapter in this book.

(2) If $|\Psi\rangle = (|0\rangle + |10\rangle)/\sqrt{2}$, then

$$\langle\Psi|a^\dagger a|\Psi\rangle = \frac{1}{2}(0 + 10) = 5 \qquad (13.197)$$

If we detect a photon, that is like applying annihilation operator to the above state. We obtain

$$a(|0\rangle + |10\rangle)/\sqrt{2} = \sqrt{5}|9\rangle \qquad (13.198)$$

This state is not normalized; we have to divide it by $\sqrt{5}$ to normalize it. Therefore the number state with nine photons is obtained at the end. The average (in fact it is exact) number of photons is nine. So, detection has increased the average number!

(3) The relevant states in this case are the ones that will "(Rabi) flop into each other":

$$|1\rangle = |g\rangle|n\rangle \qquad (13.199)$$

$$|2\rangle = |e\rangle|n - 1\rangle \qquad (13.200)$$

with the corresponding energies $E_1 = E_g + n\hbar\omega$ and $E_2 = E_e + (n - 1)\hbar\omega$. The Hamiltonian is given by

$$H = \begin{pmatrix} E_1 & V_{12} \\ V_{21} & E_2 \end{pmatrix} \qquad (13.201)$$

We obtain energies from

$$H\Psi_\pm = E_\pm \Psi_\pm \qquad (13.202)$$

The solution to this eigenvalue equation is

$$E_\pm = \frac{1}{2}(E_1 + E_2) \pm \frac{1}{2}[(E_1 - E_2)^2 + 4|V_{12}|^2]^{1/2} \qquad (13.203)$$

Large $E_1 - E_2$ leads to

$$E_\pm = \frac{1}{2}(E_1 + E_2) \pm \frac{\hbar}{2}\Delta\frac{1}{2}[1 + 2|V_{12}|^2/\Delta^2]^{1/2} \quad (13.204)$$

$$\sim \frac{1}{2}[(E_1 + E_2) \pm \hbar\Delta \pm 2|V_{12}|^2/h\Delta] \qquad (13.205)$$

The last term is the so-called Stark shift.

(4) A coherent state is the minimum uncertainty state in position and momentum of a harmonic oscillator.

This is a natural Hamiltonian from the physical perspective as it says that when a photon is lost in the field it is absorbed by the atom and vice versa.

Going to the interaction picture the Schrödinger equation reduces to

$$\hbar\lambda(\sigma_- a^\dagger + \sigma_+ a)|\psi\rangle = i\hbar\frac{\partial}{\partial t}|\psi\rangle \qquad (13.206)$$

We assume that there are n photons in the field. Then due to energy conservation only the following superposition is possible:

$$|\psi\rangle = c_1|e, n\rangle + c_2|g, n+1\rangle \qquad (13.207)$$

The Schrödinger equation becomes

$$\lambda(c_1\sqrt{n+1}|g, n+1\rangle + c_2\sqrt{n+1}|e, n\rangle) = i(\dot{c}_1|e, n\rangle + \dot{c}_2|g, n+1\rangle) \qquad (13.208)$$

Multiplying $\langle g|$ and $\langle e|$, we obtain

$$\lambda\sqrt{n+1}c_1 = i\dot{c}_2 \qquad (13.209)$$
$$\lambda\sqrt{n+1}c_2 = i\dot{c}_1 \qquad (13.210)$$

By taking the derivative of the second equation and substituting it into the first,

$$\ddot{c}_1 + (\lambda\sqrt{n+1})^2 c_1 = 0 \qquad (13.211)$$

The solution is

$$c_1(t) = A\sin(\lambda\sqrt{n+1}t) + B\cos(\lambda\sqrt{n+1}t) \qquad (13.212)$$

Therefore $\lambda_n = \lambda\sqrt{n+1}$.
But at $t = 0$, $c_1(0) = 0$, so that

$$c_1(t) = \sin(\lambda\sqrt{n+1}t) \qquad (13.213)$$

The probability is therefore

$$p_1(t) = |c_1(t)|^2 = \sin^2(\lambda\sqrt{n+1}t) \qquad (13.214)$$

If the field is in the coherent state

$$|\alpha\rangle = e^{-|\alpha|^2/2}\sum_n \frac{\alpha^n}{\sqrt{n!}}|n\rangle \qquad (13.215)$$

then the amplitude for the ground state at time t is

$$c_1(t) = e^{-|\alpha|^2/2}\sum_n \frac{\alpha^n}{\sqrt{n!}}\sin(\lambda\sqrt{n+1}t) \qquad (13.216)$$

Thus the probability is

$$p_1(t) = e^{-|\alpha|^2}\left|\sum_n \frac{\alpha^n}{\sqrt{n!}}\sin(\lambda\sqrt{n+1}t)\right|^2 \qquad (13.217)$$

If $\alpha = 0$, then

$$p_1(t) = |\sin(\lambda t)|^2 \qquad (13.218)$$

and $p_1 = 1/2$ implies $\sin(\lambda t) = 1/\sqrt{2}$, hence

$$t = \pi/4\lambda \tag{13.219}$$

(5) This is bookwork. Here is just one way of deriving it. If we divide the total time interval T into n small time intervals, then the amplitude for the transition is given by

$$\langle j|(1 - \frac{i}{\hbar}\tilde{V}(t)dt)^n|i\rangle \tag{13.220}$$

which is, up to the first order, equal to

$$\langle j|(1 - \frac{i}{\hbar}\int \tilde{V}(t)dt)|i\rangle \tag{13.221}$$

This is in the interaction picture, so converting back we have that

$$\tilde{V}(t) = V(t)e^{i\omega_{ij}t} \tag{13.222}$$

Therefore, the probability is given by the mod square of this

$$\frac{1}{\hbar^2}|\langle j|V\int e^{-i\omega t}e^{i\omega_{ij}}dt|i\rangle|^2 \tag{13.223}$$

(the positive frequency is omitted in the rotating wave approximation). Performing the integration leads to the required formula

$$P_{ij} = 4|V_{ij}|^2\frac{\sin^2((\omega_{ij} - \omega)T/2)}{(\hbar(\omega_{ij} - \omega))^2} \tag{13.224}$$

where $V = \langle j|V|i\rangle$ is the transition matrix element.

  (i) The transition from j to i has $|V_{ji}|^2$, but $V_{ji} = V_{ij}^*$ and so $|V_{ji}| = |V_{ij}|$, and this proves the equality of the rates.
  (ii) For a short time rate we have that (as can be shown by Taylor's expansion)

$$dP/dt = |V_{ij}|^2T \tag{13.225}$$

  as required.
  (iii) To obtain B, we have to average over a continuum of states (of the system or the driving field — it doesn't matter which). Once we sum up over all the states that contribute to the transition, we obtain an expression that is independent of time. This is because the integral of the above sinc function is proportional to T, so that the resulting derivative is independent of it.

(6) We know how to solve the Jaynes–Cummings model. This interaction Hamiltonian is the same apart from the fact that it involves fourth power of the creation and annihilation operators. This means population oscillations at the rate $\cos \Omega_4 t$ where $\Omega_4 = \lambda \sqrt{(n+1)n(n-1)(n-2)}$ between the state $|e, n\rangle$ and $|g, n+4\rangle$.

Bibliography

L. Allen and J. H. Eberly, *Optical Resonance and Two-Level Atoms* (New York, Dover, 1975).

D. Deutsch, *The Fabric of Reality* (London, Viking–Penguin Publishers, 1997).

B. Diu, F. Laloe and C. Cohen-Tannoudji, *Quantum Mechanics* (New York, John Wiley and Sons, 1977).

A. Einstein, B. Podolsky and N. Rosen, *Can Quantum-Mechanical Description of Physical Reality Be Considered Complete?*, Phys. Rev. **47**, 777 (1935).

H. Everett, *The Theory of the Universal Wave Function* in B. De Witt and N. Graham (eds.) *The Many-Worlds Interpretation of Quantum Mechanics* (New Jersey, Princeton University Press, 1973).

R. Feynman, R. Leighton and M. Sands, *The Feynman Lectures on Physics* (Reading, Addison–Wesley, 1989).

B. D. Guenther, *Modern Optics* (New York, John Wiley and Sons, 1990).

H. Haken, *Light II* (Amsterdam, Holland Physics Publishing, 1985).

W. Heisenberg, *The Physical Principles of Quantum Thoery* (New York, Dover, 1930).

K. Huang, *Statistical Mechanics* (New York, John Wiley and Sons, 1963).

E. T. Jaynes and F. W. Cummings, *Comparison of quantum and semiclassical radiation theories with a single mode quantized field*, Proc. IEEE **51**, 89-109 (1963).

L. Landau and E. Lifshitz, *Mechanics* (Butterworth Heinmann, 1960).

R. Loudon, *The Quantum Theory of Light* (Oxford, Oxford University Press, 1964).

L. Mandel and E. Wolf, *Optical Coherence and Quantum Optics* (Cambridge, Cambridge University Press, 1995).

C. W. Misner, K. S. Thorne and J. A. Wheeler, *Gravitation* (New York, W. H. Freeman, 1973).

N. Newbury *et. al Princeton Problems in Physics with Solutions* (Princeton, Princeton University Press, 1991).

R. Penrose, *Shadows of The Mind* (Oxford, Oxford University Press, 1993).

L. Pitaevskii and S. Stringari, *Bose–Einstein Condensation* (Oxford, Oxford University Press, 1993).

J. J. Sakurai, *Modern Quantum Mechanics* (Reading, Addison–Wesley, 1994).

E. Schrödinger, *Die gegenwärtige Situation in der Quantenmechanik (The present situation in quantum mechanics)*, Naturwissenschaften **23**, 807, 823, 844 (1935).

M. S. Scully and M. O. Zubairy, *Quantum Optics* (Cambridge, Cambridge University Press, 1997).

V. Vedral, *The role of relative entropy in quantum information theory*, Rev. Mod. Phys. **74**, 197 (2002).

J. von Neumann, *Mathematische Grundlagen der Quantenmechanic* (Berlin, Springer, 1932; English Translation, Princeton, Princeton University Press, 1955).

J. A. Wheeler and W. Zurek, *Quantum Theory and Measurement* (Princeton, Princeton University Press, 1992).

S. Weinberg, *The Quantum Theory of Fields* (Cambridge, Cambridge University Press, 1997).

F. Wilczek and A. Shapere, *Geometric Phases in Physics* (Singapore, World Scientific, 1990).

Index